Till Jan
med vänliga hälsningar
Gustav

For Tove and Agnes

Reaching for the Stars

Gustav Holmberg

Reaching for the Stars: Studies in the History of Swedish Stellar and Nebular Astronomy, 1860-1940

Lund, 1999

Holmberg, Gustav
Reaching for the Stars: Studies in the History of Swedish Stellar and Nebular Astronomy, 1860-1940

Ugglan: Lund Studies in the History of Science and Ideas, 13

History of Science and Ideas, Lund University, Lund 1999
ISBN: 91-628-3837-7
ISSN: 1102-4313

History of Science and Ideas
Dept. of Cultural Sciences
Lund University
Biskopsgatan 7
S-223 62 LUND
SWEDEN
www.kult.lu.se/ilh/

Printed in Sweden by Universitetstryckeriet, Lund

Preface

This book could not have been written without the input from many people. Even though writing a book such as this can sometimes be a lonely activity, it is very much a collaborative effort.

First of all, my supervisor Gunnar Broberg has been a source of tremendous inspiration, encouragement, kindness, and warmth. He is a scholar with a *bildning* that is out of this world. Gunnar has been a constant source of support and guidance; in retrospect I can see his stabilising presence throughout these somewhat rocky years.

Sven Widmalm in Uppsala has been my assistant supervisor. His criticism has given me new ways of looking at my texts, pointing out a lot of nonsense to be pruned down (and other parts to be expanded).

With Jakob Christensson and Jonas Hansson I have had immensely valuable discussions about being a historian. Together with Gunnar Broberg, they have been vital for my experience during my time as graduate student, as well as Svante Nordin, Sten Högnäs, Hertha Hanson, Elisabeth Mansén, Anna Tunlid and all the other people at the department. Meeting you all on a more or less daily basis provided me with an essential part of my education as a historian.

Gunnar Broberg, Sven Widmalm, Svante Nordin, Sten Högnäs, Anna Tunlid, Jan Eric Olsén, Carl Magnus Pålsson, David V. King and Tove Paulsson Holmberg came with valuable comments in the finishing phases during the hectic summer and autumn of 1999. Any remaining errors of fact or interpretation are my own.

Furthermore, I have always been very well treated at the libraries and archives I have visited. The librarians and archivists I have worked with have all given me great service.

The Lund observatory has a magnificent library where a sizeable part of the sources for this book is located. All the staff at the observatory have been most kind, but I would specifically like to mention two names: Gunnar Larsson-Leander and Gösta Lyngå, who welcomed me (as madly in love with Urania as only teenage amateur astronomers are) as a volunteer at the observatory during a couple of summers in the early eighties. For me, those summer months are a watershed in my life. I learnt a lot about astronomy, and have afterwards always felt welcome at the Lund observatory. It was also during one of those summers that I first got acquainted with Nils Dunér and historical sources, when his previously unpublished observations of the supernova of 1885 were dug out of a century of neglect, analysed, and published.[1]

I gratefully acknowledge the financial support from the Crafoord, Karin & Hjalmar Tornblad, and Fornander foundations.

I am grateful to Taylor and Francis for permission to draw from one of my previous articles for chapter two.

My parents and sister have always been very helpful and encouraging, in my academic work as well as in all other parts of life. I want to thank them for all their support and love through all these years.

I met Tove in the second semester course of history of science and ideas. Urania and Clio got competition. Tove and I soon found out that we shared very much in life; she has been by my side ever since. She has been the ideal companion for a graduate student: always there with a great mixture of love, intelligent comments and an encouraging attitude. Our lovely daughter Agnes has provided several new perspectives on life. They are the stars I reach for each day; to them, this book is dedicated.

Lund, November 1999
Gustav Holmberg

---

[1] Gerard de Vaucouleurs, Nils Hansson, Gösta Lyngå, "Dunér's Observations of SN1885 (S Andromedæ) in M31", *PASP* vol 97 (1985), 30-31.

From Classical Astronomy to Astrophysics: An Introduction

In 1938, the astronomer Bertil Lindblad addressed the whole astronomical community (or at least, those astronomers who had travelled to Stockholm to take part in the general assembly of the International Astronomical Union). IAU was by then the leading international organisation of astronomy, and its general assemblies, held every third year, were important meetingplaces for the international astronomical community. Bertil Lindblad was professor of astronomy at the Royal Academy of Sciences and director of the Academy's observatory. In his speech, Lindblad welcomed the international astronomical community to Sweden, Stockholm, and the Stockholm observatory, recently reorganised, modernised, and expanded at a site outside the city, in Saltsjöbaden. Lindblad could claim that astronomy had always been very close to the heart of the Royal Academy of Science, and that he had illustrious predecessors. "Recently transferred from its old place to a neighbouring community, the present Stockholm Observatory has still the privilege of trying to preserve, though largely active in other fields of astronomy, the traditions of prominent men of science like Wargentin, Gyldén and Bohlin."[2]

What did Lindblad mean when he claimed that his predecessors had been active in other fields of astronomy than the ones then pursued at

[2] Bertil Lindblad, "Address", *Transactions of the International Astronomical Union* vol 6 (Cambridge, 1939), 313-314, 313.

the Stockholm observatory? How can a scientific discipline like astronomy be discussed in terms of various subcultures, as Lindblad implied in his speech, and how were they related? What constituted these subdisciplinary groupings, and how did their relative strengths change over time? What ideals can be tied to the various subdisciplinary groups?

Science can be seen as built up by interactive individuals that constitute thought collectives, local cultures of scientific reasoning, and specific thought styles, to use two terms coined by Ludwik Fleck.[3] Peter Galison argues that science can be seen as a disunified activity that is broken up into several areas, and that a history of science done exclusively from the theoretical or observational perspective does not do full justice to the historical record: "the subcultures of physics are diverse and differently situated in the broader culture in which they are prosecuted. But if the reductionist picture of physics-as-theory or physics-as-observation fails by ignoring this diversity, a picture of physics as merely an assembly of isolated subcultures also falters by missing the felt inter-connectedness of physics as a discipline."[4] Instead of being a monolithical structure, it is built up of local subcultures.[5] A scientific discipline like astronomy can be seen as being built up by a number of diverse scientific cultures. Observers and theorists, classical astronomers and astrophysicists, define cultures that in some instances of history are distinct. But they are not absolute: no water-proof hulls exists between scientists from different cultures, and there are mechanisms that keep a discipline together. Such cultures of theorists, observers, and instrument-oriented astronomers are sometimes quite independent, sometimes they interact. For Galison, this "extraordinary diversity of scientific cultures that participate in the production of data" leads to a historiography of science where periodisation is not necessarily made from the standpoint of theory exclusively (or experiment).[6] Neither theory nor experiment alone is the motor of scientific progress or the basis for science. Changes in high theory do not necessarily lead to immediate changes in experimental practice; experimental/observational scientists live in a culture of their own.

[3] Ludwik Fleck, *Uppkomsten och utvecklingen av ett vetenskapligt faktum: Inledning till läran om tankestil och tankekollektiv* (Stockholm & Stehag, 1997).

[4] Peter L. Galison, *Image and Logic: A Material Culture of Microphysics* (Chicago, 1997), 46.

[5] Peter Galison, *Image and Logic*, chapter 9.

[6] Peter Galison, *Image and Logic*, 781-803, quote on 781.

Galison argues for a study of physics that does not put observation first; the faults of such a positivistic outlook have been demonstrated by decades of antipositivistic history and theory of science; observation is sometimes influenced by theory, scientific knowledge is not necessarily cumulative and empirical. But he is also critical of the prominent place taken by theory in the view of the antipositivists; empirical science does not behave like the servant of theory, it has a momentum (and an inertia) of its own.

The identification of local cultures of scientific practice does not imply that there is no communication going on, despite the many cultures of a scientific discipline. Galison has used the trading zone as a concept to describe the collaboration that goes on despite the differences in scientific cultures. Galison sees the trading zone "as a social, material, and intellectual mortar binding together the disunified traditions of experimenting, theorizing, and instrument building". It represents "an intermediate domain in which procedures could be coordinated locally even where broader meanings clashed."[7] Trading zones can also exist between scientific disciplines, as well as between scientific subcultures. I think the standpoints of Fleck and Galison indicate something important about the nature of science, as being built up of localised subdisciplinary units. One aim of the book is to discuss a number of such local scientific cultures, through a series of studies of Swedish astronomy, mainly within astrophysics and stellar astronomy.

Galison's intercalated brick model, where science is a mosaic of cultures, is akin to what this book is about. Theoretical astronomy, or the end results of astronomical enquiry, does not take centre stage; rather, I will try to give a picture of astronomy that also engages observational astronomy and the development of practice (both in observational and theoretical astronomy). Thus, it is a story of what astronomers do when they do astronomy.

One focus in this book is on scientific practice and the use of scientific technologies, specifically the recently introduced technologies of photography, spectroscopy, and the technologies connected with running a large-scale theoretical 'laboratory' in astronomy in the era before electronic computers. History of science used to be focused on the theoretical aspects of science.[8] Instead, a number of historians of science have started

[7] Peter Galison, *Image and Logic*, 803, 46.

[8] David Gooding, Trevor Pinch, & Simon Schaffer, "Preface", in David Gooding, Trevor Pinch, & Simon Schaffer, *The Uses of Experiment: Studies in the Natural Scien-*

concentrating more on scientific practice, observation and experiment. "Experimentation has a life of its own," a life that has increasingly come under historians' eyes as they explore the practice of science.[9] As Robert Smith has pointed out, historians of astronomy have for quite some time now been engaged in a study of the role played by instruments in the development of astronomy, but mainly for the time before 1900.[10]

By studying the way science is done, rather than limiting analysis to the outcome of the scientific activity, the 'result' of science, or the theories of the universe that the astronomers put forward, it is perhaps possible to further delineate the many connections between science and society, and find otherwise hidden scientific ideals not made explicit in textual accounts. It is quite unusual to find explicit accounts of scientific methodology among Swedish astronomers; for the most part, they did astronomy, rather than discussed its philosophy. Their scientific ideals will often have to be fleshed out by the historian from an analysis of how they did science. Furthermore, it is quite possible that scientific technologies, instruments, and methods for doing large empirical programmes are nuclei around which the condensation of subdisciplinary groups take place. One possibility is that the subdisciplinary identity can be seen to take place in how astronomers design instruments and weigh the relative merits of various ways of doing astronomy.

---

ces (Cambridge, 1989; 1993), xiii-xvii; Frank M. Turner, "Practicing Science: An Introduction", in Bernard Lightman ed., *Victorian Science in Context* (Chicago, 1997), 283-289.

[9] While theory-centred accounts of history of science has not disappeared, it seems that an increasing number of publications during the last decade have been written by historians interested in looking at the more practical side of science. David Gooding, Trevor Pinch & Simon Schaffer, *The Uses of Experiment: Studies in the Natural Sciences* (Cambridge, 1989; 1993); Peter L. Galison, *How Experiments End* (Chicago, 1987); Peter Galison, *Image and Logic*; Robert E. Kohler, *Lords of the Fly:* Drosophila *Genetics and the Experimental Life* (Chicago, 1994); Albert Van Helden & Thomas L. Hankins, eds., *Instruments* (*Osiris* 2nd ser vol 9 (1994)); Jed Z. Buchwald ed., *Scientific Practice: Theories and Stories of Doing Physics* (Chicago & London, 1995). While much philosophy of science has focused on representation/theory rather than on practice/experiment, Ian Hacking is one philosopher that put emphasis on scientific experimentation and claimed that experimentation has a life of its own. Ian Hacking, *Representing and Intervening: Introductory Topics in the Philosophy of Natural Science* (Cambridge, 1983; 1987), quote: 150.

[10] Robert W. Smith, "Engines of Discovery: Scientific Instruments and the History of Astronomy and Planetary Science in the United States in the Twentieth Century", *JHA* vol 28 (1997), 49-77.

Science must be seen in relation to the surrounding parts of society. Science is an activity that, in some cases, depends very much on the support of society. The connections between science and society are so many that it is almost impossible to make a distinction between 'internal' and 'external' factors. Scientific technologies are particularly an area where there is a flow of resources and ideas between science and society. Scientific technologies are an important factor in the development of science; instruments are in some instances of the history of science important for setting the research agenda. As Albert Van Helden and Thomas L. Hankins point out, instruments ought to be an important factor for the historian of science; since "instruments determine what can be done they also determine to some extent what can be thought."[11]

Thus, it is the aim of this book to look in some detail into a number of scientific technologies used by Swedish astronomers during the late nineteenth and early twentieth centuries. What uses were made of the technologies, what scientific ideals were tied to various instruments and methods of observing the sky and handling observational data? There is also a practice aspect to theoretical astronomy, and it is the aim of this book to look at practice in a number of settings, observational as well as theoretical.

*

During the eighteenth and early nineteenth centuries, the stars were primarily a backdrop that defined a frame in which the main action of *classical astronomy* occurred: the motions of the sun, moon, asteroids and planets. Astronomers did astrometry to observe the positions of celestial bodies. These observations were then used for the work in celestial mechanics, that used Newtonian mechanics to study the past and future motions of the celestial bodies.[12]

Observationally, classical astronomy was about precise measurements (of angles and time) that gave information on the positions of celestial bodies; the laws of Newtonian mechanics gave certainty to the theoretical picture. There was a self-image of exactness; measurements and calcu-

---

[11] Albert Van Helden & Thomas L. Hankins, "Introduction: Instruments in the History of Science", *Osiris* 2nd ser vol 9 (1994), 1-6, 4.

[12] Introductions to the classical era are found in Dieter B. Herrmann, *The History of Astronomy from Herschel to Hertzsprung* (Cambridge, 1984); Michael A. Hoskin, *The Cambridge Illustrated History of Astronomy* (Cambridge, 1997).

lations were precise. Planets were discovered – Uranus in 1781, Neptune in 1846 – as well as an increasing number of minor planets and comets. These could be fitted in to the system by applying the methods of celestial mechanics on the data presented by the astrometrical observations. Theory and observation fitted together.

Beyond the limits of the solar system, an early attempt by William Herschel to delineate the stellar world resulted in the distribution of the stars in a lens-shaped system. Herschel could not measure the distances to the stars, but his idea was that the system's diameter was about 850 Sirius distances.[13] The first measurements of stellar parallaxes came in the 1830's, but still the majority of stars were at unknown distances. A handful of stars could be reached by the parallactic method by 1850, but most astronomers nonetheless worked with the motions of celestial objects. The nature of these bodies and the structure of the universe were areas that astronomers did not emphasise.

The emergence of a new astronomy brought change to the science. A new collection of astronomical problems and methods were developed; the way of doing astronomy changed. Astrophysics was born (the field was also called solar physics, the new astronomy, cosmical physics, astronomical spectroscopy). The astronomical methods of astrophysics made it possible to discuss questions like the composition and physical state of a distant celestial object, and the development of the stars.[14]

A number of astronomical technologies were important for the emergence of astrophysics during the second half of the nineteenth century. *Spectroscopy* changed what astronomers could work with. With this method, an astronomer could determine what substances were present in a distant celestial body. The classification of stellar spectra eventually led to various hypotheses about stellar evolution. Spectroscopy meant that the astronomer imported several ways of doing science from physics. Spectroscopy radically altered the working methods of the astronomer, who had to learn new skills.[15] *Photography* also changed the way astronomers worked. The real breakthrough for photography as an astronomical tool came during the 1880's, after the introduction of the dry gelatine plate.

---

[13] Herrmann, *History of Astronomy*, 11ff.

[14] For introductions to the emergence of astrophysics see Owen Gingerich ed., *Astrophysics and Twentieth-Century Astronomy to 1950* (*The General History of Astronomy* vol 4A). (Cambridge, 1984).

[15] Cf. Simon Schaffer, "Where Experiments End: Tabletop Trials in Victorian Astronomy", in Jed Z. Buchwald ed., *Scientific Practice*, 257-299.

Photography entailed a mechanisation of vision and an industrialisation of observation: many stars could be captured on one plate which was measured for data on stellar positions and brightness during office hours; the photographic plate was considered, by many astronomers, to be a more precise detector than the human retina. The photographic technologies were used both in classical astronomy as well as in stellar astronomy and astrophysics. Given the increased mobility of photograpic plates, photography also contributed to a division of labour in astronomy; some observatories concentrated on photographic observations, and the plates could be transported to astronomers specialising in the reduction and analysis of photographic plates. There was also a division of labour within an observatory. Reduction of positions of stars or classification of spectra on plates could be done by others than professional astronomers; the introduction of (often female) assistants digging for data on the plates was a technology for getting more data, often employed at several observatories. Photography changed the very nature of being an astronomer. The third technology mentioned here, *photometry*, dealt with the measurements of the brightness of celestial objects. These measurements could be done both visually and photographically. The photometrical reduction of photographic plates became an important part of photographic astronomy, and for many astronomers a thorough knowledge and understanding of the chemistry and physics of the photographic process became a vital part of the scientific toolkit.[16]

The new astronomy meant in part a spatial expansion; more astronomers dealt with phenomena of individual stars, the structure of the system of stars, and the structure of the cosmos beyond our stellar system, rather than with the solar system. Stellar and nebular astronomy were areas that attracted increasing numbers of astronomers in the period around 1900. Stellar statistics was a way of trying to get at the structure of the stellar system by using statistical methods on the large amounts of data collected with photographic observation. The stellar statisticians arrived at models of the stellar system that were on the order of 10 000

---

[16] John B. Hearnshaw, *The Analysis of Starlight: One Hundred and Fifty Years of Astronomical Spectroscopy* (Cambridge, 1986); John Lankford, "The Impact of Photography on Astronomy", in Owen Gingerich ed., *Astrophysics and Twentieth-Century Astronomy to 1950*; John Hearnshaw, *The Measurement of Starlight: Two Centuries of Astronomical Photometry* (Cambridge, 1996); Klaus Staubermann, *Controlling Vision: The Photometry of Karl Friedrich Zöllner* (unpublished PhD thesis, Darwin College, Cambridge, 1998).

light-years with the sun placed near the centre. Through the work by Harlow Shapley and others, many astronomers came to accept a larger system during the 1920's, one in which the stellar system had a diameter of between 100 000 and 300 000 light-years, with the sun placed far away from the centre. It was an enormous restructuring of humanity's place in the universe. The status of the nebulæ also came under debate. Proponents of the island universe theory stated that the nebulæ were distant collections of stars on a scale similar to our stellar system; other astronomers argued that the nebulæ were local objects in our stellar system, perhaps being stars under formation from the contraction of clouds of dust. Measurements by Adriaan van Maanen of the rapid rotation of several spiral nebulæ were interpreted as signs of these objects being local; such fast angular rotation periods would translate to absurd velocities if the objects were distant assemblies of large numbers of stars, rather than local objects of gas and dust inside our stellar system. Several astronomers, with Edwin Hubble as the most famous, got results around 1925 that convinced astronomers that many nebulæ were distant systems of stars. Following observations by Hubble and theoretical work by Lemaître and others, the expansion of the universe was discovered.[17] Thus, the period from about 1860 to 1940 is a dynamic period in the history of astronomy.

It is important to keep in mind that classical astronomy by no means disappeared from observatories during the early parts of the twentieth century. Astronomers continued to do celestial mechanics, astrometry, or observe phenomena on the planetary surfaces. The new astronomy was amended to classical astronomy; astronomy was broadened. Classical astronomy and the new astronomy co-existed within the discipline of astronomy, but it is probably fair to say that the relative strength of classical astronomy diminished during the first decades of the century.[18]

*

[17] Erich Robert Paul, *The Milky Way Galaxy and Statistical Cosmology, 1890-1924* (Cambridge, 1993); Robert W. Smith, *The Expanding Universe: Astronomy's 'Great Debate' 1900-1931* (Cambridge, 1982); Gudrun Wolfschmidt, *Milchstrasse, Nebel, Galaxien: Strukturen in Kosmos von Herschel bis Hubble* (München, 1995).

[18] Karl Hufbauer and Ronald Doel have made this point about the conditions in American astronomy, and it is probably applicable to several other astronomical communities. Ronald Doel, *Solar System Astronomy in America: Communities, Patronage, and Interdisciplinary Science, 1920-1960* (Cambridge, 1996), 16.

Stellar astronomy became something of a specialty for Swedish astronomers during the time period covered here, at least in the latter part. An increasingly large number of astronomers began to specialise in stellar astronomy. The relative number of astronomers that did astrometry and celestial mechanics diminished compared to stellar astronomy. Stellar astronomy was, in turn, made up of a number of subfields, for example stellar statistics, stellar spectroscopy and photometry, etc. A main task for the stellar astronomers was to solve the problem of how stars were arranged in space. The stellar statistical school used a statistical treatment of empirical data to attack the problem, whereas later a group formed that used spectroscopy to measure distances to many stars. Together with a number of astronomers that studied nebulæ, stellar astronomy (especially with an empirical style) became a leading part of Swedish astronomy during the first decades of the twentieth century, which is mirrored in the selection of subjects for these studies. The majority of these deals with astronomers that in one way or another studied stellar astrophysics, the distribution of stars in space, the structure of our stellar system and other stellar systems. Special emphasis is put on a number of observatory directors. What were their styles, their scientific personalities, and how did they influence the scientific practice at their observatories?

Around the middle of the nineteenth century, astronomy in Sweden was pursued at three institutes: Lund University, Uppsala University, and the Stockholm observatory, tied to the Royal Academy of Science. Astronomy at Uppsala had a brand new observatory, constructed between 1844 and 1848; in 1853 the director of the observatory, Gustaf Svanberg moved in there (in those days, it was not unusual to find astronomers living at observatories). In 1860, the observatory was furnished with its main instrument, a refractor made by Steinheil of München.[19] At Lund, the astronomers dwelled in the tower of Kungshuset, the old university building. A new observatory was soon constructed also in Lund: in 1867 observations at the new Lund observatory started. The main instrument here was a ten inch refractor, supplemented in 1874 by a meridian circle, a type of instrument that was constructed for meas-

[19] Carl Schalén, Nils Hansson, Arvid Leide, *Astronomiska observatoriet vid Lunds universitet* (*Ur Lunds universitets historia*, 4) (Lund, 1968); Gunnar Malmquist, "Hur Uppsala fick sitt nuvarande observatorium – ett hundraårsminne", *PAT* vol 35 (1954); Erik Holmberg, "Lundensisk astronomi under ett sekel", *Cassiopeia: Tycho Brahe sällskapets årsbok* 1949, 15-54; yearly reports of the Lund, Stockholm and Uppsala observatories in *LUÅ*, *UUÅ*, *ÖKVAF*.

uring the positions of stars. At Stockholm, astronomers worked at the observatory dating from 1753. The Stockholm astronomers had the responsibility of calculating and publishing the Swedish almanac, the publication of which was monopolized by the Royal Swedish Academy of Sciences. Astronomy and astronomers had also been working with geodetic surveys and the mapping of the land.[20]

Astronomy could be said to have fairly good resources by 1870; two out of three observatories were newly constructed and furnished with modern instruments. Astronomers worked almost exclusively in the classical tradition: celestial mechanics, astrometry, time keeping. This dissertation is mainly focused on the emergence of astrophysics, stellar and nebular astronomy, but it is important to keep in mind that the tradition of classical astronomy was very strong in Sweden. In the first phase, Nils Dunér and other astronomers connected with the new way of doing astronomy attempted to introduce the new astronomy alongside the classical astronomy. Gradually, the new astronomy was gaining ground and by the end of this period, classical astronomy was a minor part of Swedish astronomy. The new astronomy was at first seen as a splinter group by some astronomers, but in the end astronomy was transformed from the inside; no separate astrophysical disciplines or institutes formed alongside the traditional observatories; the astrophysicists rather changed astronomy within existing institutional borders. This process of change is one theme here. To delineate the process, a number of studies are made.

From the 1870's, several Swedish astronomers started working with photography and spectroscopy. At Lund, Nils Dunér investigated the spectra of red stars and spectroscopically determined the solar rotation. At Stockholm observatory, Hugo Gyldén began work in photographic astronomy. When Dunér got a chair at Uppsala, he continued to develop photographic and spectroscopic astronomy there. Gradually, photography and spectroscopy became part of the practice of Swedish observatories. Photography was often discussed in terms of objectivity, especially when the technology was presented to other people than astronomers. The mechanical nature of the photographic observations was seen as something that entailed objectivity. The use of photography changed the way astronomers worked; the observations were industrialised. Photography was not the sole property of astrophysics, but it was also adapted to classical fields like astrometry. The introduction of photography and

[20] Sven Widmalm, *Mellan kartan och verkligheten: Geodesi och kartläggning, 1695-1860* (Uppsala, 1990).

spectroscopy and the use of these new technologies is the theme of chapter two.

In chapter three, the work of Carl Vilhelm Ludvig Charlier is discussed. Here, an attempt is made to study the work of theoretical astronomers in a practice-oriented way. After becoming professor at Lund observatory in 1897, Charlier developed an activity in stellar statistics that hinged on the statistical handling of large amounts of data. For this he set up a large-scale operation consisting of computers – in those days, a computer was a human doing routine calculations – equipped with mechanical calculating machines. Charlier and his group developed statistical methods which was subsequently used in other sciences. Charlier and his pupils also often worked in areas of society that were far away from astronomy but needed to handle large amounts of data. Thus, several astronomers offered their statistical expertise to companies and the state.

In chapter four, Knut Lundmark and the activities at Lund observatory under his leadership are studied. Lundmark worked on nebular astronomy and participated in the process of changing the large-scale picture of the cosmos in the 1920's. Later at the Lund observatory, he began a programme of assembling a large body of observational data on nebulæ into a catalogue. The project failed, and Lundmark became more and more marginalised in Swedish astronomy. This process of marginalisation happened simultaneously as Lundmark increased his writing of popular astronomy and became widely known as an author of popular astronomy.

Chapter five focuses on a group of astronomers formed around Bertil Lindblad at Uppsala and Stockholm is discussed. Lindblad and his group worked on the distribution of stars in the Milky Way, the motions of the stellar system, problems of absorption of light by interstellar matter and so on. Lindblad and his group achieved international recognition for their work. The 1938 general assembly of the IAU was held at Stockholm. They had a strong position in Swedish astronomy. The new Stockholm observatory, placed at Saltsjöbaden and inaugurated in 1931, was geared to the work of unravelling the Milky Way. Methods were developed there for measuring the distances to the stars from low dispersion spectra, following a method deviced by Lindblad in the early 1920's. Lindblad also worked on the motion of stars in the Milky Way; together with the Dutch astronomer Jan Oort, he analysed the motion of stars in the rotation of the Milky Way system.

In chapter six, a number of professorial appointments are studied to analyse the arrangement of power within the discipline. Experts' statements show how arguments were made about the relative merits of the various ways of doing astronomy. In this chapter, research in fields close to astronomy are also discussed, as are the international contacts of Swedish astronomy. To provide some perspectives on the science of astronomy, some activities in popularisation and history of astronomy are discussed.

*

A wholly different book could of course be written, with focus on other aspects of Swedish astronomy: classical astronomy, popularisations of astronomy, astronomers and the history of astronomy, astronomy-related themes in fields close to astronomy such as geophysics. A number of such themes are discussed rather briefly here and in chapter six.

Several programmes were pursued in observational classical astronomy. For example, the Lund observatory participated in the Astronomische Gesellschaft's large programme to determine positions of stars with a meridian circle. Lund did the zone between +35 and +40 degrees declination (analoguous to latitude on the sky). The work started in 1878 and was published between 1895 and 1900.[21] Later, during the 1920's, Walter Gyllenberg at Lund reobserved the Lund zone with the same instrument. A similar programme for observing stellar position with the meridian circle ran at the Stockholm observatory. Astronomers also observed the positions of minor planets and comets visually, by measuring the positions relative to stars with the help of a micrometer attached to a refractor.

When it comes to the theoretical side of classical astronomy, the work of Hugo Gyldén (1841-1896) was central to Swedish astronomy.[22] Gyldén became director of the Stockholm observatory in 1871 after having worked at the Helsingfors (where he was born) and Pulkovo observatories. Gyldén specialised in celestial mechanics. He worked on methods for calculating the orbits of comets, minor planets, and planets. During the 1880's he published several works on the orbits in the solar system. Gyldén was seen as a leading astronomer in celestial mechanics, both in

---

[21] Carl Schalén, Nils Hansson & Arvid Leide, *Astronomiska observatoriet vid Lunds universitet* (*Ur Lunds universitets historia*, 4) (Lund, 1968). 74ff.

[22] Bertil Lindblad, "Gyldén, Johan August Hugo", *SBL*.

Sweden and abroad. In 1884 he was called to a professorship at Göttingen but was convinced to stay in Sweden. A fund that allowed financing his teaching at the Stockholm Högskola was organised (king Oscar II made the largest contribution to the fund).[23] He was president of the Astronomische Gesellschaft between 1889 and 1896.

In 1874, his introductory book *Framställning af astronomin i dess historiska utveckling och på dess nuvarande ståndpunkt* was published. It is wholly centred on classical astronomy. Stellar astronomy and astrophysics are not discussed; in the introduction, Gyldén makes it clear that he figures these areas belong outside of astronomy.[24] Astronomy is defined as the science of the laws for the motions of celestial objects. Practical astronomy is the determination of the places of such objects by observation, whereas theoretical astronomy is the determination of the real motion from observations of the apparent positions; physical astronomy is the calculation of a celestial body's motion from a knowledge of mechanical laws and the forces that are acting on the body. In recent times, however, another kind of investigation had begun to take place, concerning the physical state of the celestial bodies. These, Gyldén argued, lack the mathematically rigorous methods used in astronomy. The knowledge gained from these newer ways of observing the sky is useless for astronomy proper. For Gyldén, these newer astrophysical methods "will probably in a near future make up a new, independent science. Therefore they will not be treated at all here."[25]

Something of a school formed around Gyldén. Several astronomers were influenced by his methods, and during the 1880's and 1890's, Sweden had several prominent astronomers working in celestial mechanics, among them Anders Lindstedt and Karl Bohlin. His wife Therèse had far-reaching cultural interests, and Hugo Gyldén played the piano well. Scientists and artists often gathered in their home, and helped make Hugo Gyldén something of a centre for Stockholm people interested in science and culture.

Theoretical astronomy existed mainly within classical astronomy. Stellar astronomy and astrophysics were different. Observational astronomy was generally stronger within stellar astronomy. One exception was

---

[23] The Stockholm *Högskola* (university college) was a complement to the Lund and Uppsala universities. Inaugurated in 1878, it specialised in the sciences.

[24] Hugo Gyldén, *Framställning af astronomin i dess historiska utveckling och på dess nuvarande ståndpunkt* (Stockholm, 1874), 1-6.

[25] Hugo Gyldén, *Framställning av astronomin*, 6.

Charlier's programme in stellar statistics. Another exception to this pattern was the work of Hugo von Zeipel at Uppsala observatory. von Zeipel had initially been observator, a post that was later transferred to a personal professorship. Working initially with celestial mechanics, von Zeipel also took up stellar astronomy and astrophysics, and he came to see himself as being mainly a theoretical astrophysicist.[26] He studied the distribution of stars in clusters of stars to try to determine the dynamics of clusters and the masses of stars.[27] In the late 1920's, he studied the theory of stellar constitution and evolution. For this work he got the Morrison award from the New York Academy of Sciences.[28] von Zeipel taught advanced courses for students at Uppsala on topics in theoretical astrophysics such as the energy levels of the hydrogen atom, the relativistic equation for an electron in an electromagnetic field, the stellar models of Eddington, the stellar energy problem etc.[29]

*

As for the history of Swedish astronomy, Sven Widmalm has studied astronomy up until 1860 in *Mellan kartan och verkligheten*, with emphasis on geodetic surveying. N.V.E. Nordenmark has published mainly on the history of astronomy in Sweden prior to the mid-nineteenth century. For the period since the 1860's, Anita Sundman has published *Den befriade himlen*, a biography of Knut Lundmark, based on the Lundmark archive. Per Collinder has published *Swedish astronomers 1477-1900.* Several astronomers have published histories of their observatories.[30] Popular astro-

---

[26] von Zeipel tried to further the cause of theoretical astrophysics by lobbying for Eddington for a Nobel prize in physics. He argued that Eddington's theoretical work on the structure of stellar interiors had benefited physics. Hugo von Zeipel to Elis Strömgren 3 January 1924 [!, 1925 is the correct year], Strömgren papers, LUB.

[27] Hugo von Zeipel to Elis Strömgren 20 May 1916, 3 January 1921, Strömgren papers, LUB.

[28] Hugo von Zeipel to Elis Strömgren 8 September 1929, 22 October 1930, 10 January 1931, Strömgren papers, LUB.

[29] Journals of the mathematical scientific section of the faculty of philosophy, spring semester 1930, autumn semester, UUB.

[30] Per Collinder, *Swedish astronomers 1477-1900* (Uppsala, 1970). Carl Schalén, Nils Hansson & Arvid Leide, Astronomiska observatoriet; Knut Lundmark, *Astronomien i Lund 1667-1936: Glimtar och drag ur astronomiska institutionens liv och öden i nuvarande och gången tid* (Lund, 1937); Erik Holmberg, "Lundensisk astronomi under ett sekel", *Cassiopeia: Tycho Brahe sällskapets årsbok* 1949, 15-54; Gunnar Malmquist, "Hur Uppsala fick sitt nuvarande observatorium - ett hundraårsminne", *PAT* vol 35

nomy in Sweden during the nineteenth and twentieth centuries will be the topic of a forthcoming study by Johan Kärnfelt. Kjell Jonsson has studied the popularisation of astronomy by Knut Lundmark and others.[31]

Gunnar Eriksson's *Kartläggarna* is a useful introduction to Swedish science of the period. In a forthcoming book, Sven Widmalm studies physics, centred on the Ångström group at Uppsala. For twentieth century physics in Sweden, see the collection of studies edited by Svante Lindqvist, *Center on the Periphery*, and Thomas Kaiserfeld, *Vetenskap och karriär*. Elisabeth Crawford has published a biography of Svante Arrhenius. Hjördis Kristenson's *Vetenskapens byggnader under 1800-talet* is a study of nineteenth century observatories and other scientific buildings in Sweden.[32]

(1954). The buildings at Saltsjöbaden observatory is the topic of Ilse-Beth Byström, "Stockholms observatorium i Saltsjöbaden: Ett byggnadsverk av Axel Anderberg", *AT* vol 15, 152-163.

[31] Kjell Jonsson, "Naturvetenskap, världsåskådning och metafysiskt patos i mellankrigstidens Sverige", *Lychnos* 1992, 103-146; "Physics as Culture: Science and Weltanschauung in Inter-War Sweden", in Svante Lindqvist, ed., *Center on the Periphery: Historical Aspects of 20th-Century Swedish Physics* (Canton, MA, 1993)

[32] Sven Widmalm, *Mellan kartan och verkligheten*. For other Swedish sciences during the period, see, for example, Gunnar Eriksson, *Kartläggarna: Naturvetenskapens tillväxt och tillämpningar i det industriella genombrottets Sverige 1870-1914* (Umeå, 1978); Svante Lindqvist ed., *Center on the Periphery*; Thomas Kaiserfeld, *Vetenskap och karriär: Svenska fysiker som lektorer, akademiker och industriforskare under 1900-talets första hälft* (Lund, 1997); Sven Widmalm, *Det öppna laboratoriet: Fysikhistoriska fallstudier, 1853-1910* (forthcoming). Elisabeth Crawford, *Arrhenius: From Ionic Theory to the Greenhous Effect* (Canton, MA, 1996); Hjördis Kristenson, *Vetenskapens byggnader under 1800-talet: Lund och Europa* (Lund, 1990).

New Technologies, new Astronomy

Around 1850, astronomy was a precision-oriented science dealing with the positions and motions of heavenly bodies. Observational astronomers worked with observing and reducing observations; they were interested in obtaining the positions of heavenly bodies to high levels of accuracy. Theoretical astronomers worked with tying these observations together, using Newtonian mechanics.

The emergence of astrophysics and photographic astronomy in Sweden is the subject here. The new practice of astrophysics transformed astronomy. Practical astronomy was not homogeneous after the middle of the century. Observational astronomy breaks up into at least two subcultures. There was classical astronomy, which observationally dealt with astrometry, precise observations of the positions of celestial objects by instruments like the meridian circle or the micrometer attached to a refractor. Astrophysics, the other main culture, was primarily dealing with the observational study of the spectra of celestial objects. The cultures had different ties with theory. Tied to classical astronomy was celestial mechanics that had been developed for many years and was based on Newtonian mechanics. Spectroscopy had not, in its earlier phases, such a firm rooting in theory. It dealt more with classification and centred more around visual arguments than the numerically-centred classical astronomy. Visual and classification work was important, but a precise and quantified method was also present in astrophysics and spectroscopy, like the measurement of wavelengths of spectral lines and measurements of the Doppler effect. Thus, astrophysics could – like classical astronomy – develop an ideal of numerical work; at least, that was the case in Sweden.

Nils Dunér

A key figure in the introduction of photography, spectroscopy, and astrophysics in Swedish astronomy was Nils Dunér (1839-1914).[33] A minister's son, he moved from Billeberga to study at the university at Lund in 1855. Besides astronomy, he studied chemistry, physics, and mathematics. His thesis was in classical astronomy; in May 1862 Dunér defended his dissertation on the orbit of the asteroid Panopea. Dunér became assistant (*amanuens*) at the Lund observatory some months later, working under Professor Axel Möller.[34]

Early in his career, Dunér participated in the polar expeditions that were an important part of science in Sweden at the time. The expeditions allowed scientists from different fields to work together.[35] Dunér was geographer and physicist in the 1861 expedition to Svalbard, led by Otto Torell and financed by Prince Oskar, the Royal Academy of Sciences and a number of private citizens. Dunér travelled to the northern and western parts of the group of islands, where he made astronomical observations to determine geographical positions; he also measured the inclination of the magnetic field. One task, reconnaissance of the possibility of an expedition for measuring a meridian arc, had to be postponed.

In 1864, Dunér participated in an expedition led by Adolf Erik Nordenskiöld, financed by the government. One goal of this expedition was to advance the plans for a future expedition to measure a degree of meridian. Dunér and Nordenskiöld published two works that put the geographical knowledge of Svalbard on a more secure basis. The astronomical observations led to better determinations of geographical positions.[36]

---

[33] Östen Bergstrand, "Nils Christofer Dunér", *SBL*; Hugo von Zeipel, "Nils Christofer Dunér", *KVAÅ* 1916, 290-312.

[34] The hierarchy of positions at observatories in Sweden was almost constant during the period studied in this thesis. The head of the observatory was the professor; under which worked the observator, a kind of assistant professor. Under the professor and observator, there were one or several assistant astronomers (*amanuens*). In several cases in this period, the observator post was made into a personal professorship.

[35] Cf. Urban Wråkberg, *Vetenskapens vikingatåg: perspektiv på svensk polarforskning 1860-1930* (Stockholm, 1995); Gösta Liljequist, *High Latitudes: A History of Swedish Polar Travels and Research* (Stockholm, 1993).

[36] Nils Dunér & A.E. Nordenskiöld, "Anteckningar till Spetsbergens geografi", *KVAH* vol 6 no 5 (1865); Nils Dunér & A.E. Nordenskiöld, "Förberedande under-

Dunér and Nordenskiöld patriotically argued for Sweden's role in the exploration of the arctic. Here, they followed on a long tradition that identified a mission for Swedish science: to pursue a scientific exploration of the arctic.[37]

These expeditions belonged to a kind of scientific enterprise that was on the rise in Sweden. During the next four or five decades, Sweden's polar expeditions allowed (often young) scientists to participate in an activity that brought them large datasets for later analysis; it also brought themselves and science substantial amounts of popularity. The scientific traveller, observing the aurora under the northern sky or at the risk of his life hacking his way in to some northern rock to expose minerals and fossils, was a character that fascinated the public. Patriotic heroism was thus also a factor in how these expeditions come to take place and be seen.

These expeditions were part of a wider theme in nineteenth century science. The fact-collecting missions on a large geographical scale that often tried to understand the laws of nature by surveying data in large scale, was a vital field of science. Alexander von Humboldt made scientific travel popular, and during the second half of the century, Swedish scientists embarked on several projects of scientific expeditions, part of the large programme of Humboldtian science. Dunér travelled together with scientists from other fields. These expeditions were cross-disciplinary scientific milieux, which for a period of some months or a year was temporarily set up in the arctic wilderness. These borealic academies gave the participating scientists the chance to work in a setting that was well financed. Back home, the datasets produced by the expeditions could be utilised to do science for years.[38]

But let us return to Lund. In 1864 Dunér became *observator* at the Lund observatory. At this time the observatory was about to enter a transitional phase. Placed in a tower in Kungshuset, the astronomers didn't have a department building of their own; modern instrumentation was also lacking. For quite some time, the professor of astronomy John

---

sökningar rörande utförbarheten af en gradmätning på Spetsbergen", *KVAH* vol 6 no 8 (1866).

[37] Sven Widmalm, *Mellan kartan och verkligheten*, 398.

[38] Susan Faye Cannon, *Science in Culture: The Early Victorian Period* (New York, 1978), chapter three: "Humboldtian Science". Gunnar Eriksson has identified such a spirit of surveying that pervaded Swedish science during the late nineteenth and early twentieth centuries. Gunnar Eriksson, *Kartläggarna*.

Mortimer Agardh had argued for funds for a new building and instrumentation; now such funds were made available by the parliament. In 1867 the astronomers in Lund moved to the new institute.[39]

The main instruments in the new observatory were a refractor of 24.5 cm aperture and a meridian circle (added later, in 1874). The refractor's mechanical parts were made by Edvard Jünger of Copenhagen. Jünger was a supplier of equipment to the Copenhagen university, the Danish military, and scientists abroad. He delivered the mechanical parts of the refractor to the new observatory at Östervold in Copenhagen, inaugurated in 1861.[40]

Dunér first tried to observe nebulæ with the refractor. For about two weeks he surveyed 15 zones on the sky with meagre results.[41] Dunér instead began a program of double star observation with the refractor, an activity that he worked on for several years.[42] In 1876, Dunér's work on double stars was published, after he had made some 2300 observations of 452 pairs.[43]

Dunér next began observing the spectra of stars and thus became part of a trend which thrived in several places around the world: astrophysics. Dunér changed the settings of his astronomical work, by becoming a member of the group of astronomers working in astrophysics. But he did this not by moving to another institute; he used a tactic and a practice that enabled him to make a change within the existing institutional setting by rearranging and introducing new auxiliary instruments.

---

[39] Carl Schalén, Nils Hansson, Arvid Leide, *Astronomiska observatoriet vid Lunds universitet* (*ur Lunds universitets historia: 4*) (Lund 1968), 44-56.

[40] C.A. Clemmensen, "Jünger, Frederik Gottlieb Edvard", *Dansk biografisk leksikon* 2nd ed. (Copenhagen, 1981).

[41] The records of the aborted project are in Nils Dunér, "Genommönstring af nora himlen med afseende på Töckenfläckar", the observatory's journals, LUB. From September 25 to October 4 1867 Dunér searched for new nebulæ, finding only a few.

[42] For calibrating the instrument, Dunér observed artificial double stars placed on the church tower of Uppåkra, about 4 km:s away. Schalén, Hansson, Leide, 72ff. During a visit by the king Karl XV, it is said that a picture of the king himself was placed on the distant church tower and observed with the refractor, to show the immense power of the instrument. *Lundahorizonten* no 2 (1938), 6.

[43] Nils Dunér, *Mesures micrométriques d'étoiles doubles, faites a l'observatoire de Lund, suivies de motes sur leurs mouvements relatifs* (Lund, 1876).

Spectral Classification, Solar Rotation, and the Technology to Measure it

In the mid-1870s, astrophysics and spectroscopical studies of astronomical objects were not very new. Several astrophysicists abroad - the Hugginses, Angelo Secchi, H.L. d'Arrest, Hermann Carl Vogel and Oswald Lohse - worked on the spectroscopical study of celestial objects; in Germany the planning of the Potsdam astrophysical observatory was already on its way in 1871.[44] In Sweden, physicists Anders Ångström and Robert Thalén worked on the spectroscopy of the sun; their work was more on the principles of spectroscopy and wavelength determinations than in astronomy.[45] Closer to Lund and Dunér, the Danish astronomer Heinrich Louis d'Arrest worked in Copenhagen on the spectra of stars; at the time of his death in 1875, he had published four papers on the spectra of red stars.[46]

Several authors mention d'Arrest as an inspiration for Dunér's migration to the field of stellar spectroscopy.[47] The source of this information is not mentioned, but it might very well have come from Dunér himself. Lund was close to Copenhagen and astronomers from Lund and Copenhagen met frequently, in part because the refractor had much Danish technology.[48] The refractor's mechanical parts were made by E. Jünger and C.V. Holten; Dunér and Möller travelled to Copenhagen now and then to have the instruments adjusted. Contacts were frequent and it seems obvious that Dunér should know of his Danish colleague's work on the spectra of red stars.

The portrayal of Dunér's entry into stellar spectroscopy as a result of inspiration from d'Arrest is probably not the whole picture. When Dunér mounted a spectroscope on the Lund refractor for the first time in 1878, stellar spectroscopy had been a rather well-known field of astronomical enquiry for years. For an astronomer who had exhausted his in-

---

[44] John B. Hearnshaw, *The Analysis of Starlight*, chapter 4; Dieter B. Herrmann, "Hermann Carl Vogel", *DSB*; Dieter B. Herrmann, "Potsdam Astrophysical Observatory", *GHA* vol 4A, 130ff.

[45] Sven Widmalm, *Det öppna laboratoriet*.

[46] Sally H. Dieke, "Heinrich Louis d'Arrest", *DSB*.

[47] Hugo von Zeipel, "Nils Christofer Dunér", 294; Schalén, Hansson, Leide, *Astronomiska observatoriet*, 81.

[48] See "Dagbok för Lunds Observatorium 1868 6/11 - 1875 22/8", the observatory's papers, LUB.

terest and instrumental means in classical astronomy and was looking for something new, stellar spectroscopy was an opportunity, ready for being taken up; this is an important background to Dunér's decision, rather than inspiration from one person, from d'Arrest. It was also a step that was not wholly safe, institutionally. Some astronomers argued that the kind of astronomy Dunér started doing did not belong to astronomy proper.[49] Thus, a career in astrophysics and stellar astronomy could become a path away from central fields of astronomy. Another possibility of course was that the new astronomy would open up a new field and become central in its own right within astronomy.

Dunér started out with a spectroscope made by the maker Heustreu of Kiel. The design for this instrument, bought by Dunér in March 1878, was made by a leading German stellar spectroscopist, Vogel, and the choice was made after a recommendation by Vogel.[50] Another and more flexible spectroscope was bought by Dunér in November of 1879, made by the Munich maker Merz. It was fitted with three different setups of prisms, for varying the dispersion, the amount of spectral detail visible. A micrometer was fitted to the eyepiece which allowed measurements of the wavelengths of spectral lines. A third instrument, from Schmidt and Haensch of Berlin was bought in 1883; it had a dispersion between that of the other two.[51]

The work of Dunér with these spectroscopes was quite close to the work of Vogel, who had been working on the classification of stellar spectra for some years. In 1873 Vogel and Lohse had started a work that aimed at classifying all stars between -10 and +20 degrees of declination and brighter than magnitude 4.5. In classifying the stars, they made use of a system of three classes. The first class was white stars with few and faint spectral lines, class two was yellow stars with more lines – like the sun's spectrum – and class III was made up of red stars with broad spectral bands of absorption. For Vogel, these classes represented stars in varying degrees of evolution, a notion that Dunér also took up.[52] He concentrated

---

[49] Hugo Gyldén, *Framställning af astronomin i dess historiska utveckling och på dess nuvarande ståndpunkt* (Stockholm, 1874), 1-6.

[50] Vogel to Dunér January 19 1878, Dunér correspondence, LUB.

[51] Nils Dunér, "Sur les étoiles a spectres de la troisième classe", *KVAH* vol 21 (1884) no 2, 11.

[52] Hearnshaw, *Analysis*, 81. For more on the classification systems, see David H. DeVorkin, "A Sense of Community in Astrophysics: Adopting a System of Spectral Classification", *Isis* vol 71 (1981), 29-49; Hearnshaw, *Analysis* ch. 4 and 5.

on the spectra of red stars, since these were in an advanced evolutionary state.[53]

Several hundreds of these red stars of class III were known from the work of other observers. Dunér made it his task to observe and classify them according to Vogel's scheme. Vogel had given Dunér advice on how to use the classification system.[54] Observing several hundred of these stars with the Lund refractor and the spectroscopes took some years. By 1884 Dunér's catalogue of red stars was published.

In these times, astronomical spectroscopy and astrophysics was in a classifying stage. Classification was an important part of the astrophysicists' work. Early stellar spectroscopy was using classification as a means to get knowledge of stars. Theoretical studies generally came later on. Before theoretical work on a large scale is undertaken, it is important for scientists to establish the empirical material, bring order and frame nature. This is often done by making atlases that establish a set of typical phenomena in the varying nature.[55] Classification programmes can also be thought of as establishing such standardised sets of nature.

The importance of international communication in the introduction of a scientific technology is suggested by the fact that Dunér got his spectroscopic technology from Germany. When a new scientific technology is introduced in a field, several components migrate to new institutional milieux; there is a transport of technology as such - of instruments and other machinery - and there is also a movement of more or less articulated ideas about that technology. The *hardware* of the technology came from the German instrument makers Heustreu, Merz and Schmidt & Haensch, while the *software* came from d'Arrest and Vogel. The fact that Dunér doesn't seem to have received any help from Thalén in Uppsala at this stage suggests that while astrophysics did thrive in the area between astronomy and physics, in this case the Uppsala physicists primarily were seen as physicists, not astronomers. They worked mainly in the field of wavelength determination, and were interested in fundamental issues like the constitution of matter.[56]

Having published his catalogue of stellar spectra, Dunér began to pursue solar spectroscopy. The main interest of Dunér was to get precise

---

[53] Dunér, "les étoiles", 6.

[54] Vogel to Dunér 12 August 1878, Dunér correspondence (film copy), LUB.

[55] Lorraine Daston & Peter Galison, "The Image of Objectivity", *Representations* vol 40 (1992), 81-128

[56] Sven Widmalm, *Det öppna laboratoriet.*

measurements of the solar rotation, by measuring the shift of spectral lines because of the Doppler effect. In the grant application for the instrument needed for this work, Dunér argued that he was about to prove a law of nature. This proof was to come about by increasing the precision of the measurements. Measurements of the Doppler effect caused by the sun's rotation had been made earlier, so Dunér's goal was to get even more accurate measurements.[57] If the solar rotation as measured by the Doppler effect in the solar spectrum came out the same as that inferred from observations of sunspots, the nature of the Doppler effect as a law of nature could be proved. When he applied for money from the foundation Lars Hierta's Minne, Dunér argued that he was about to prove a law of nature by measuring the Doppler effect caused by the solar rotation.[58] Dunér got the money and initiated the construction of a spectroscope so powerful that it could measure the minute shift in spectral lines because of the Doppler effect stemming from the solar rotation.

Dunér had got good advice from Vogel when he started out in stellar spectroscopy. A similar role was played by Bernhard Hasselberg (1848-1922), a Swedish physicist and astronomer who had worked at the Pulkovo observatory outside St Petersburg since 1872, after studies at the Uppsala university. At Pulkovo, Hasselberg had got the chance to use more advanced equipment than any available in Sweden at the time. In 1881 Dunér took up correspondence with Hasselberg in Pulkovo. Hasselberg had come into contact with an "exquisite" grating made by Henry Rowland.[59] Hasselberg was mainly interested in using the Pulkovo grating for doing a revision of the wavelengths of the lines in the solar spectrum. Dunér queried Hasselberg about the merits of gratings in general as compared to prisms, and Rowland's gratings in particular. Hasselberg's results showed that gratings made measurements of very high precision possible; he recommended that Dunér get a grating made by Rowland. He also gave Dunér valuable tips on how to construct the spectroscope; for example using the same lens for collimation and observation, an arrangement that would bring down the size of the instrument. An instrument with this kind of resolution was bound to be quite large, and

[57] Hearnshaw, *Analysis*, 148ff.

[58] Dunér to Hasselberg 18 November 1887, UUB okat. 466 i:1.

[59] Hasselberg to Dunér 7 December 1883, Dunér correspondence (film copy), LUB.

bringing down the physical size was important. Otherwise it might not fit the refractor or make the telescope mechanically unstable.[60]

Dunér did give the prism alternative a thought – by writing the British instrument maker Adam Hilger and asking about the possibilities of a high dispersion prism spectroscope – but settled for a grating spectroscope. He asked Rowland, who agreed on selling him a grating.[61] Henry Rowland, professor of physics at the Johns Hopkins university, had constructed a device that made it possible to make high precision gratings, ruled on mirror metal. The number of gratings that was produced was not high and it was not always easy to get access to these gratings. Rowland, however, sold Dunér two high quality gratings.

The mechanical parts of the spectroscope were ordered from Christopher Peter Jürgensen in Denmark. Jürgensen had taken over the firm of Edvard Jünger, maker of the mechanical parts of the Lund observatory refractor. Jürgensen's firm was a supplier of advanced technology for the military, science, navigation, office equipment, surveying etc. They worked on the electricity system of Copenhagen, delivered torpedo equipment to the navies of Sweden and Norway, made an early typewriter etc.[62]

For his work on the solar spectrum, Dunér used the high dispersion instrument for measuring the small Doppler shift of the sun's light from the limbs of the sun. Dunér compared the positions of the lines with the telluric stationary lines caused by oxygen in the earth's atmosphere. The effect is small, on the order of 0.04 Å.[63] Earlier measurements of solar ro-

---

[60] Hasselberg to Dunér 1 February, 20 October and 6 December 1884, Dunér correspondence (film copy), LUB; Dunér to Hasselberg 26 January 1884, UUB okat. 466 i:1.

[61] Adam Hilger to Dunér 29 January 1884, Dunér correspondence (film copy), LUB.

[62] C.A. Clemmensen, "Christopher Peter Jürgensen", *Dansk biografisk leksikon* 2nd edition, (Copenhagen, 1981).

[63] I have observed the effect visually using a fairly modern spectroscope on a solar telescope at the Lund observatory. When the mirror in that telescope is swung from the eastern to the western limb, the lines in the solar spectrum jump a bit, compared to the telluric lines that stand still; the effect is very small, even in a modern instrument constructed specifically for solar observations. That Dunér could get good results by measuring this small effect is an indication of his skill as an observer, and the power of his spectroscope. (Incidentally, the lens in this modern solar telescope, used in teaching of solar astronomy at the Lund observatory, is from the old refractor, installed in 1867 and cannibalized when that telescope was dismantled. Sometimes old

tation had timed the passage of sunspots across the solar disk. If the rotation time of the sun as measured with the Doppler shift was consistent with such observations of sunspots, the theory of the Doppler shift could be said to be vindicated.[64]

The work by Dunér on the solar spectrum at Lund was published in 1891.[65] The results agreed well with data from sunspots. Some years later, Dunér repeated the solar rotation work at Uppsala.[66]

The solar observations were visual and quantitative. Dunér measured the positions of iron lines relative to oxygen lines produced in the atmosphere of the earth with a micrometer. When it came to precise measurements of line positions, photographic technologies still did not reign supreme.[67]

Astrophysics and the spectroscopy of astronomical objects show how important auxiliary instrumentation is for the history of astronomy. Telescopes represent a more slow-moving astronomical technology. To participate in the new astronomy, it was not a new telescope Dunér primarily needed. The refractor at Lund was of rather modest size when it was constructed in the 1860's; its size was even more modest compared to state of the art telescopes ten or twenty years later. Dunér however could take part in astrophysical front-line research by adding proper auxiliary instrumentation to this telescope. The use of auxiliary instrumentation is thus an important part of astronomical technology. Changes in that kind of instrumentation have often changed the way astronomers work. Proper use of mobile auxiliary instrumentation could also be a way around a situation where it was not institutionally possible to devote an instrument exclusively to one observational tradition.

For Dunér it was possible to modernise the instruments at the Lund observatory by retrofitting the refractor with auxiliary instrumentation. The spectroscopes that were employed on the telescope made it possible

---

instruments or parts of them live on; they show up in new configurations in the mosaic of scientific instruments.)

[64] John Hearnshaw, *Analysis*, 148f.

[65] Nils Dunér, *Recherches sur la rotation du Soleil* (Uppsala, 1891).

[66] Nils Dunér, *Über die Rotation der Sonne. Zweite Abhandlung.* (*Nova acta regiæ Societatis scientiarum Upsaliensis* ser 4 vol 1 no 6 (Uppsala, 1907).

[67] Cf a statement by the Harvard physicists John Trowbridge and Wallace Clement Sabine in 1887. They argued that when it came to wavelength determination, visual observation with a micrometer still was better than photography, which lent itself more to qualitative mapping of the spectrum. Klaus Hentschel, "Photographic Mapping of the Solar Spectrum 1864-1900, Part I", *JHA* vol 30 (1999), 93-119,110.

for him to partake in the new astronomy, and be part of the astrophysical community. It was often talk about the auxiliary instrumentation that was discussed in letters between Dunér and his astrophysical colleagues. Auxiliary instruments added to a telescope that belonged to the classical era of astronomy made it possible for Dunér to work in the emerging new field of astrophysics. It was the resource Dunér needed to establish the telescope in another field of astronomy. Dunér's work in getting the proper tools for astrophysics enabled him to move towards astrophysics and, in effect, transport the Lund refractor to a whole other field of astronomy. This occurred in an institute were the director was not an astrophysicist but active in classical astronomy; hence, Dunér could not count on changing the identity of the telescope permanently by changing the whole telescopic outfit.[68]

Photography and Spectroscopy at Uppsala Observatory

Dunér was seen in Sweden as a major astronomer by the late 1880's: his work on the spectroscopy of the red stars and the sun were at the forefront of the new astronomy, astrophysics. He had also received international recognition for his work in astrophysics. In 1887 the Institut de France awarded him the Lalande price.[69] When the professorship at Uppsala became vacant after Herman Schultz retired he was called as professor, rather than having to apply for the post. He now had the chance to participate in the modernisation of the observatory, and steer the institute in a direction that he wanted. Photographic astronomy was a vital part of the modernisation set in motion by Dunér.

The introduction of the photographic plate to astronomy was delayed. Initially astronomers expressed excitement as to the promises of this technology, but the high hopes sometimes gave way to frustration. Still, the use of photographic technology did, after some decades, become an

---

[68] By adding the spectroscope to the refractor, the tower at the Lund observatory became a site for the production of astrophysical data. The use of auxiliary instruments provided a means for having the Lund observatory jumping between two diverse scientific cultures, that of classical astronomy and that of astrophysics. Once Dunér had access to knowledge about spectroscopic practice and the proper spectroscopical tools, he could start work at temporarily transforming the refractor into a machine for producing astrophysical data.

[69] Östen Bergstrand, "Dunér, Nils Christofer", 532.

important part of the astronomer's way to observe the sky. Especially after the introduction of the dry gelatine plate, astronomers increasingly used photographic technologies. With the dry gelatine plate, photographic material became easier to handle and, with the increased sensitivity amd the ability to expose plates for longer periods of time, better equipped to reach fainter objects. In this section, the introduction and later use of photographic technologies is discussed.

Bernhard Hasselberg once again appears in the introduction of new astronomical technologies by way of his working at Pulkovo. In 1873, soon after having joined the Pulkovo observatory staff, Hasselberg started to learn the new photographic and spectroscopic technologies. His first teacher in photography was a geodesist connected with the Russian military, Stubendorff. He also corresponded with the Czech Weinek on photographic techniques such as recipes for developers.[70] The immediate reason for this interest was the forthcoming transit of Venus in 1874, which the Pulkovo astronomers observed photographically, like astronomers from other institutes.[71]

The next step was to institutionalise the new astronomy as a department within the Pulkovo observatory. During the academic year ending in May 1877, an astrophysical laboratory was built up in Pulkovo. One reason for the observatory's interest in astrophysics was stated in the first paragraph of the report of the opening of the lab: spectroscopy made it possible to observe radial velocities of stars. Work on photographic issues was also to be performed at the laboratory; the newly observed transit of Venus had raised issues about photography, for instance the amount of exposure time that was suitable in solar photography. It was also stated that Hasselberg wanted to use photography applied to spectroscopy to determine to what extent mixture of various elements influenced the spectra.[72]

Hasselberg embarked on a study of the spectra of various elements. He was interested in measuring the spectra of hydrocarbons, since such lines were visible in the spectra of comets. In the first years work was focused

[70] Bernhard Hasselberg, "Sjelfbiografiska anteckningar", vol 1, p 167, UUB okat. 466 i:3; Weinek to Hasselberg 18 November 1873, 16 January 1874, UUB okat. 466 i:3. (The first letter is stained with spots from what could be photographic chemicals.)

[71] John Lankford, "The Impact of Photography on Astronomy", *GHA* vol 4A, 23.

[72] *Jahresbericht am 11. Mai 1877 dem Comité der Nicolai-Hauptsternwarte abgestattet vom Director der Sternwarte* (St Petersburg, 1877), 8ff.

on laboratory spectroscopy, his observations were not then performed on the spectra of celestial objects.[73]

The astrophysical laboratory also handled photographic work on the sun. When the photographic equipment that had been used in the transit of Venus expedition to Vladivostok had been restored to working order, a series of solar photographs was started. The aim was to produce a material that allowed a study of the formation and disappearance of sunspots.[74]

The name given to the department implies a connection between astronomy as an observational and an experimental science; both kinds of work were done there.[75] Hasselberg even made a study of the observatory's lightning conductors in 1881-2; he found that these did not function properly, and managed to put them back in working order.[76]

There are several letters in Hasselberg's correspondence from these years on photographic matters. The astronomers gave each other tips on photographic plates, recipes for developers, designs for cameras and so on.[77] Hasselberg's activities in the new areas of astronomical practice, astrophysics, spectroscopy, photography, were not without friction from other astronomers. Otto and Hermann Struve were not always sympathetic with Hasselberg's interests in the new astronomy.[78]

Hasselberg's knowledge of the nuts and bolts of the new scientific technologies earned at Pulkovo was useful to Swedish astronomers as they were about to modernise their way of observing the sky. Hasselberg's work at the Pulkovo laboratory was a parallel line to the introduction of new practices in Swedish astronomy. The Russian observatory thus played an important role in the history of Swedish astronomy. Dunér is important in the introduction of photographic and spectro-

---

[73] *Jahresbericht am 20. Mai 1878 dem Comité der Nicolai-Hauptsternwarte abgestattet vom Director der Sternwarte* (St Petersburg, 1878), 9f; *Jahresbericht für 1878-79 und 1879-80 am 24. Mai 1880 dem Comité der Nicolai-Hauptsternwarte abgestattet vom Director der Sternwarte* (St Petersburg, 1880), 18ff.

[74] *Jahresbericht am 20. Mai 1881 dem Comité der Nicolai-Hauptsternwarte abgestattet vom Director der Sternwarte* (St Petersburg, 1881), 12f.

[75] Simon Schaffer, "Where Experiments End".

[76] *Jahresbericht am 19.Mai 1882 dem Comité der Nicolai-Hauptsternwarte abgestattet vom Director der Sternwarte* (St Petersburg, 1882), 20f.

[77] See letters from, for example, Asaph Hall, J.H. Chrisite, J.M. Eder, Eugen von Gothard, William Huggins, Oswald Lohse to Hasselberg in UUB okat. 466 i:1, and Hermann Carl Vogel, Edward Pickering and Weinek in UUB okat. 466 i:3.

[78] Hasselberg to Dunér 16 February 1886, Dunér correspondence (film copy), LUB.

scopic technologies. When he became professor of astronomy at Uppsala, his skills in these observational technologies were useful as he tried to steer the institute there towards new ways of doing astronomy.

When Dunér arrived at Uppsala the department was, just like the observatory in Lund, geared towards the classical mode of observational astronomy.[79] The main instrument at Uppsala was a refractor of about the same size as the one in Lund, 24 cm:s. It had been installed in 1860 and was corrected for visual observation; it belonged to the visual age of astronomy, and was virtually useless for photography.[80]

Funding became available for Dunér to introduce the new technologies at Uppsala. Dunér now had the ability to acquire an instrument that was from the outset more geared towards modern astronomy, instead of pragmatically changing the capabilities of an existing instrument and shifting such an instrument's mode of operation to another type of astronomy by adding the proper auxiliary instrumentation. "When will the old men [*gubbarna*] in the second chamber [of parliament] grant funds for an astrophysical institute?" Bernhard Hasselberg wrote to Dunér in a letter where he praised the advancing astrophysical research at Pulkovo.[81] Now that had, in a sense, come by. Instead of having an instrument of classical astronomy temporarily made into an astrophysical instrument by changing the surrounding material culture, Dunér now could have a group of instruments made from the outset into astrophysical instruments. But this array of astrophysical apparatus was still present at an astronomical observatory that also contained instruments for work in classical astronomy and astronomers working in the classical field. The emergence of astrophysics did not lead to the birth of a separate astrophysical institute. Astrophysics grew within the existing discipline of astronomy. Classical astronomy did not disappear overnight, but its dominance of the science was challenged.

One could differentiate between the instrument at large and the instrument in detail. By the instrument in detail I mean a particular in-

---

[79] Dunér had tried to introduce photographic astronomy in Lund during the 1880's but failed, partly because his professor, Möller, could not get the necessary funds. Dunér to Hasselberg 18 November 1887, UUB okat. 466 i:1.

[80] The lenses of refractors also act as prisms, dispersing the white light into colours. By using different kinds of glass for two or more lenses in the telescope, the refractor was corrected for this chromatic aberration. Usually, correction was possible only for parts of the spectrum; the visually corrected telescopes could not be used for photography with the blue-sensitive plates.

[81] Hasselberg to Dunér 1 February 1884, Dunér correspondence (film copy), LUB.

strument: the telescope, a spectroscope, a photographic camera etc. The instrument at large is the whole assembly of such individual instruments used by the astronomer. It takes some institutional engineering to bring such an instrument at large into action. In his previous activity at the Lund observatory, Dunér had to construct anew such an instrument at large for astrophysical observations from a collection of classical and modern instruments in detail. For Dunér at Uppsala, it was possible to get an instrument at large to operate by fitting pieces together into an assembly with a goal from the outset. Dunér could construct the instrument at large with more degrees of freedom than he had at the Lund observatory.

His vision was to get "a new refractor, suitable for all the methods of observations of the new astronomy".[82] The instrument type chosen by Dunér for the observatory at Uppsala was the double refractor, consisting of two telescopes mounted in tandem, one with visual and one with photographic colour correction. This kind of telescope was common during a period when the visual way of observing the sky had not yet become obsolete. Dunér did not want to "part with the possibility of seeing the sky, and there are situations, especially with variable weather, when photography still could not supersede the direct observations."[83] The new instruments should enable the astronomer to pursue both astrophysics and precision measurements of double stars.[84] The instrument was referred to as "a large astrophysical refractor [...] equipped with micrometer, finder, spectroscope, polarising helioscope, and other auxiliary instrumentation that belongs to larger refractors outfitted for astrophysical work."[85] It was a modern instrument. Among other things, the illumination of the scales and micrometer filaments was to be electrical, not oil-burning lamps as before. The electrification of astronomical instrumentation is a marker of the new age astronomers were working in.

The double refractor for the Uppsala observatory came with a 36 cm visual and a 33 cm photographic tube. The lenses came from Steinheil of Munich, the mechanical parts from Repsold of Hamburg. In October 1892, Dunér travelled to Germany. In Berlin and Potsdam he discussed possible spectroscopic and electrical equipment for the new instrument. In Hamburg, he picked up the delicate lenses, which in two crates, with a

[82] Dunér, yearly report of the Uppsala observatory, UUÅ 1889, 55.
[83] Dunér to Hasselberg 1 March 1886, UUB okat. 466 i:1 (Dunér's emphasis).
[84] Dunér to Hasselberg 23 July 1890, UUB okat. 466 i:1.
[85] Uppsala observatory yearly report, *UUÅ* 1891, 57.

combined weight of 72 kilograms, were transported "via Berlin, Lübeck, Malmö and Stockholm to Upsala, in the railway cars and steamboat cabins in which the director himself travelled; when travelling through these cities, he never lifted his eyes off them".[86] The instrument was assembled the last days of December 1892, after the two crates containing the mechanical parts and totalling over 2000 kilograms, had arrived. In 1892, as if to crown the institutional arrival of astrophysics with further international recognition, the Royal Society awarded Dunér the prestigious Rumford medal in gold for his work in astrophysics.[87]

The Uppsala refractor was designed in such a way that it allowed Dunér and his fellow astronomers to pursue both classical astronomy and astrophysics. The double nature of the double refractor allowed Dunér to work in what he called precision astronomy - measurements of binary stars and other kinds of astrometry - and astrophysics. It also had a flexibility in that it allowed astronomers to observe both visually and photographically. The instrument was constructed at a time when photography and visual observation often were used in conjunction; in some cases, astronomers preferred visual observations, in others the photographic.[88] Thus it was important to have an instrument that catered to both modes of observation. It was an instrument that different subcultures of astronomy could use. It was a balance point between the classical and the new astronomy. It mirrored Dunér's career - he had worked in both classical astronomy and astrophysics - and it mirrored the state of astronomy in the latter decades of the nineteenth century; astrophysics had not become dominant, and substantial numbers of astronomers still worked on classical research programmes. The discipline of astronomy had not split up entirely with the advent of astrophysics - classicists and astrophysicists often worked in the same departments. Therefore, an instrument was designed that catered to both groups. In terms of astrophysics and photographic astronomy in Sweden, it was a step forward. While in Lund, Dunér had made the refractor into an astrophysical instrument by adding auxiliary instrumentation to an older instrument. At Uppsala, Dunér could put the new astronomy - both astrophysics and photographic astronomy - on an equal footing with classical astronomy; they were incorporated into the very design of the instrument and the surrounding workspace at the observatory.

---

[86] Uppsala observatory yearly report, *UUÅ* 1893, 70f.

[87] Östen Bergstrand, "Dunér, Nils Christofer", 532.

[88] Klaus Hentschel, "Photographic Mapping", 109f.

The new techniques for observing the sky were integrated into the new machinery at the observatory. The refractor had spectroscopes and it was outfitted for photographic astronomy. A photographic laboratory was integrated into the observatory milieu. A machine for precise measurements of stellar positions on the photographic plates was eventually installed. Dunér and his fellow astronomers and astrophysicists celebrated the arrival of a modern photographic equipment by arranging a *Diner Astrophotographique* at the Uppsala observatory.[89]

It was not only the astrophysicist Dunér who used photography; several other Swedish astronomers where using the new technology for observing the sky, even though it was the Uppsala observatory, under Dunérs leadership, that acquired the most extensive set of photographic telescope technology. At Stockholm observatory, Hugo Gyldén was professor of astronomy at the Royal Academy of Sciences. While his interests as an astronomer was focused on celestial mechanics, he also started doing photography. In the summer of 1887 he wrote Dunér that he had "begun to dabble in photography as this way of observation soon will make a complete revolution in practical astronomy".[90] The weather was, however, not that good; observing was more successful during the following year, when plates taken with a 8 cm astrographic telescope, clamped to the 17 cm telescope, showed stars to the 13th magnitude. The material was used by C.V.L. Charlier, assistant astronomer (amanuens) at the observatory for discussing the use of photographic plates for photometry, the measurement of the magnitudes of stars.[91]

Lund observatory was to be the last to be equipped with photographic equipment. When C.V.L. Charlier became professor in 1897 the observatory got photographic equipment. Charlier ordered an astrograph of 15 cm aperture.[92]

The Carte du Ciel was a project that grew out of a suggestion by E.B. Mouchez of the Paris observatory. He had been impressed by the results

[89] The menu from the dinner is preserved, written by Dunér. The wines: "Bains de Développment Pyrogallol de Margaux Méthol de Heres" etc [Dunér's emphasis]. Hasselberg papers, UUB okat. 466 i:1.

[90] Gyldén to Dunér 27 July 1887, 28 July 1887, 24 December 1888, Dunér correspondence (film copy), LUB.

[91] Yearly report of the Stockholm observatory's activities, *ÖKVAF* 1889, 216f; Gyldén to Dunér 24 Decemberg 1888, Dunér correspondence (film copy), LUB; C.V.L. Charlier, *Über die Anwendung der Sternphotographie zu Helligkeitsmessungen der Sterne* (Publ. der Astron. Gesselleschaft, no 19) (Leipzig, 1889).

[92] Lund observatory yearly reports, *LUÅ* 1897, 1898.

of the Henry brothers, working on photography at the Paris observatory. Mouchez envisioned a project were copies of the Henry telescope were to be used at observatories all over the world to photograph the sky. Plates were to be measured; a catalogue of stellar positions and an atlas were to be published.[93]

In April 1887 astronomers met in Paris to discuss Mouchez' proposal. Hugo Gyldén and Nils Dunér came from Sweden and Hasselberg from Pulkovo. Dunér and Hasselberg were sceptical as to the project; they wanted to hold the project until better photographic technologies had been invented. They pointed to the difference in colour sensitivity between the plates and the human eye; yellow and red stars that were prominent to the eye, showed up fainter, relative to blue stars, on the photographic plates.[94] Previous astronomical observations, based on visual methods, could not easily be transparently compared with photographic observations.

The cost for an observatory participating in the Carte project was high, and Dunér was critical of Swedish participation in such a project; it would block resources that could be used for astrophysics and other kinds of research.[95]

Both astrophysics and classical astronomy were involved in the introduction of photography in Swedish astronomy. While the most extensive use of photography was by the astrophysicist Dunér, classical astronomers like Gyldén and Charlier also used the new technology. It was seen as a new way to observe the sky both for astrophysics and classical astronomy.

Uppsala was the major observatory in terms of photographic observations in Sweden at the turn of the century. When the asteroid Eros made a close passage in 1900-1901, Dunér made photographic observations of it, as part of an international campaign to observe Eros and deduce the

---

[93] John Lankford, "Impact of Photography", 27ff.

[94] Nils Dunér, "Om den på fotografisk väg framställda stjernkatalogen", *ÖKVAF* 1900, 399-407, 405; Dunér to Hasselberg 11 March 1887, UUB okat. 466 i:1; von Zeipel, "Dunér", 305.

[95] Dunér to Hasselberg 23 July 1890, UUB okat. 466 i:1. Thirtyfive years later, with the Carte du Ciel project still unfinished, a discussion about having Uppsala take over a part of the Carte du Ciel was turned down. Uppsala observatory did not have the necessary funding, and such a project would be a strong contraint on the work possible to undertake at the observatory, its director, Östen Bergstrand, argued. Östen Bergstrand to Knut Lundmark 29 June, 17 September, 1925, Lundmark's collection, LUB.

solar parallax. Otherwise, he continued with visual observations, following up with new observations of old projects; the spectra of red stars, and the rotation of the sun. Dunér had constructed a modern observatory with photographic technologies, but it was mainly his students that used it. Dunér's pupil, Östen Bergstrand, started working on photographic determination of stellar parallaxes in 1894.[96] Bergstrand discussed the technology for measuring parallaxes, and measured the parallaxes of several stars with the photographic refractor.

Around 1907, Bergstrand began to shift his research interests, moving towards more astrophysical applications of photography; he started measuring the colours of stars photographically. These observations were initially performed with the same instrument; the double refractor was possible to use by both kinds of astronomical practice. Bergstrand worked out a method that involved placing a coarse diffraction grating made out of metal rods or a mesh in front of the lens of the telescope. The coarse grating produced very low dispersion spectra of stars on the plate. The distance between the practically pointlike first-order spectra on plates taken through such gratings is proportional to the maximum of energy in a star's spectrum, the so-called effective wavelength, and gives information on the temperature of the star. The distribution of energy in stellar spectra thus could be reduced to the measurement of a distance between two points on a photographic plate; this way of observing gave new astrophysical data of stars but also had connections with the precision oriented use of photography Bergstrand had practiced in his determinations of parallaxes. The method was an outcome of the local scientific mileu devised by Dunér, which combined an ethos of precise measurement with modern astrophysical observation.

Ejnar Hertzsprung and Östen Bergstrand used the method early on; later, Bergstrand's pupils Bertil Lindblad and Knut Lundmark used it to measure the colours of many stars.[97] In his correspondence with Hertzsprung, Bergstrand learned useful hints on how to observe effective wavelengths, and what types of astrographs (photographic telescopes), plates,

---

[96] Knut Lundmark, "Carl Östen Emanuel Bergstrand", *SBL*.

[97] C. Davidson, "Astrophotographie", *Handbuch der Wissenschaftlichen und angewandten Photographie*, vol 6 part 1 (Vienna, 1931), 102-233, 157ff; Simon Newcomb and Rudolf Engelmann, *Populäre Astronomie* 6th ed (Leipzig, 1921), 544. See also David DeVorkin, "Stellar Evolution and the Origin of the Hertzsprung-Russell Diagram", Owen Gingerich ed., *The General History of Astronomy* vol 4A, 90-118; Axel Nielsen, "Contributions to the History of the Hertzsprung-Russell Diagram", *Centaurus* vol 9 (1964), 219-253.

and auxiliary equipment were suitable. In a sense, Bergstrand became Hertzsprung's temporary student in the fields of photographic colorimetry and photometry, picking up a knowledge of the practice through an exchange of technical tips. Hertzsprung provided software for the technology.[98]

However, for these observations Bergstrand wanted new hardware. The photographic telescope ordered by Dunér in the early 1890's had a small field of view because of the long focal length which gave a large plate scale. An instrument with a wider colour correction was also desirable; as in all refractors, light of all colours was not brought to a common focus, and with different designs of the instruments, better colour correction could be achieved. An opportunity for Bergstrand to get funds for a new instrument came with the solar eclipse of August 1914, visible from the northern parts of Sweden. Bergstrand's strategy for getting funds was that he emphasized the rarity of this phenomenon, and the need for effective instruments to observe it. Outwardly, he argued that the observatory needed an instrument to study the rare phenomenon of a solar eclipse that happened to be visible from within the borders of Sweden, whereas Bergstrand actually wanted an instrument that was suitable for his new orientation of photographic astronomy, the observations of stellar temperatures through measurement of the effective wavelength. He managed to obtain funds for a double astrograph with 6-inch lenses, made by Zeiss of Jena. The mechanical parts were constructed by Heide of Dresden.[99]

Bergstrand continued to pursue the field of effective wavelength during and after the war. He developed the technology and tried to organise an international scheme of standardisation to calibrate the method. This programme attracted initially the interest of several observatories.[100]

---

[98] Se letters from Hertzsprung to Bergstrand 11 January 1909, 19-26 August 1912, 23 April 1913, 7 May 1913, 28 May 1913, 13 June 1913, 16-17 June 1913. Bergstrand's collection, NC:736, UUB.

[99] Hertszprung to Bergstrand 19-26 August 1912, Bergstrand's collection, NC:736, UUB; yearly report of the Uppsala observatory, *UUÅ* 1913, 131f.

[100] F.W. Dyson to Östen Bergstrand 12 December 1922, UUB; Östen Bergstrand & H. Rosenberg, "Vorschlag zur Aufstellung einer Normalsequens zur Bestimmung effektiver Wellenlängen", *AN* vol 215 (1922), 447-452.

The Solar Eclipse in 1914

At the turn of the century, astronomers no longer had to observe phenomena such as prominences and the spectrum of the chromosphere at total solar eclipses. Rather, the main interest for sending out eclipse expeditions was the tenuous solar atmosphere, the corona, which still could not be observed outside totality, despite attempts by George Ellery Hale and others. Success came in the early 1930's, when Bernard Lyot successfully constructed an instrument for observing the corona outside of eclipse.[101] Questions on the physical nature of the corona, structural changes, and so on had to be investigated during a few minutes of totality a few times each decade, and from locations scattered throughout the world. This was a task well suited for the photographic plate: with some quick exposures material was collected that was analysed after the eclipse. Photography enabled astronomers to stop the time, as it were, and analyse the observations at a later time.[102] The photographic observations were, not surprisingly, hailed by the astronomers as a vital step forward toward a more objective way of acquiring observations and knowledge of the solar corona than the previously used technique: drawing. The photographic observations of the form of the corona were considered to be of a higher quality than the visual ones, which had been "subjective"; the photographic observations, by contrast, "being objective representations of reality, at once achieved the character and value of exact observations", Bernhard Hasselberg claimed.[103]

A total eclipse of the sun was visible from northern Sweden on 21 August 1914. Preparations for observing the eclipse were made under the auspices of the Royal Academy of Sciences, the initiative having been taken by Vilhelm Carlheim-Gyllensköld.[104] A commission for co-ordinating

---

[101] Karl Hufbauer, "Artificial Eclipses: Bernard Lyot and the Coronagraph", *HSPS* vol 24 part 2 (1994), 337-394.

[102] Photographic methods had gradually been taken into use at eclipses during the period 1870-1890. Alex Soojung-Kim Pang, "The Social Event of the Season: Solar Eclipse Expeditions and Victorian Culture", *Isis* vol 84 (1993), 252-277, 270ff.

[103] Bernhard Hasselberg's account of the solar eclipse expedition of the Royal Academy of Sciences, *KVAÅ* 1915, 99-125, 101.

[104] Carlheim-Gyllensköld, once assistant at the Stockholm observatory, was in 1914 a physicist at the Academy of Sciences and the Stockholm University College. For more on Carlheim-Gyllensköld and astronomy, see Anders Carlsson & Gustav Holmberg, "Vilhelm Carlheim-Gyllensköld på Stockholms observatorium", *Lychnos* 1995, 179-189, and references therein.

observations was organised with Nils Dunér, Bernhard Hasselberg, Karl Bohlin, C.V.L. Charlier, Svante Arrhenius, Gustaf Granquist, Carlheim-Gyllenskcöld and Östen Bergstrand (the elderly Dunér soon withdrew). A proposal to Parliament for funds resulted in a grant of 60 000 kr.[105] A programme with meteorological, magnetic, and electric measurements was drawn up, to supplement the astronomical observations, which consisted of contact timings and photographic observations of various kinds.[106] The spectacular event made it possible to buy instruments that could be used for other observations too. In fact, all three Swedish observatories acquired new photographic equipment for the eclipse.

From Stockholm observatory, an expedition was sent to Långsele, where Bohlin took spectrograms of the corona using a spectrograph on a 10-inch Zeiss reflector, bought especially for the eclipse. With a smaller instrument pictures were taken of the outer parts of the corona.[107] Lund observatory had an expedition stationed at Strömsund, where a new 7-inch astrograph was used to search for possible intra-Mercurial planets. Uppsala's expedition was at Österforse, where Bergstrand used the new Zeiss-Heyde double astrograph for photographing the corona. The plates were later measured with a microphotometer to obtain measurements of the distribution of light in the corona.[108]

The largest of the instruments studying the solar eclipse (and one of the largest cameras ever in Sweden) was used by the Royal Academy of Sciences' Physics Department's expedition to Sollefteå. Led by Bernhard Hasselberg, its aim was to observe the corona in very high resolution. The horizontal camera's objective had a focal length of 20 meters and was fed by a coelostat, an arrangement of clock-driven mirrors that compensated for the Sun's motion across the sky. The optical parts were manufactured by Zeiss, whereas the mechanical parts were constructed by Gus-

---

[105] Royal Academy of Sciences minutes 22 May 1912, KVA. There were tensions in the committee, because of competition for the money allotted by parliament. See, for example, Carlheim-Gyllensköld to Charlier 11 March 1913, Charlier's collection, LUB.

[106] Proposal to the Royal Academy of Sciences for a commission to plan work at the solar eclipse of 1914, signed by Carlheim-Gyllensköld, 22 May 1912, and minutes at the meeting of the commission 20 July 1912. Hasselberg's collection, UUB okat 438f:1, UUB.

[107] Report of the Stockholm observatory eclipse expedition, *KVAÅ* 1915, 95-97.

[108] Östen Bergstrand, *Études sur la distribution de la lumière dans la couronne solaire* (L'éclipse totale de soleil des 20-21 août 1914, deuxième partie, no 2) (Stockholm, 1919).

taf Rose, the academy's instrument-maker. Hasselberg ordered plates measuring 70 x 70 and 50 x 50 cm:s from three makers of photographic materials: Agfa of Berlin, Leto Photo-materials of London, and Union photographique industrielle of Lyon. The French plates got held up in Antwerp because of the outbreak of the First World War, and arrived after the eclipse.

A hut contained the cassettes for the plates, as well as Hasselberg and assistants. They communicated with personnel at the other end of the huge camera by telephone. Four plates were exposed by the observers who worked in a mechanical way: "These operations had during the days preceding the eclipse been thoroughly trained and were executed during the totality with a fully automatic precision. Not a word could be uttered, or the whole operation might well have got into hopeless disorder".[109] The concept of mechanical objectivity pertained also to the observers, who needed to attain a mechanical performance during the short totality, unless things would become mixed up. In a sense, the observers were part of the machine. Hasselberg was inside the camera hut during the eclipse, and hence could only see a short glimpse of the magnificent phenomenon on the camera's ground glass between exposures. The photographic plate lay between the observer and the solar eclipse, and the astronomer had become part of the machine. Hasselberg was perhaps somewhat disappointed with this but in the hunt for scientific results such aesthetic reasonings had to be put aside.[110]

Hasselberg's plates showed coronal structures in great detail, but no scientfic analysis was published by him.[111] One reason is that another, greater darkness fell upon Europe in August 1914. Hasselberg wanted to study the changes in coronal structure over a small time-scale by comparing his plates with plates taken by a German expedition to the Crimea. Because of the outbreak of war, that expedition fell foul of the Russian authorities, and the comparative study was made impossible.[112] An expedition from the Lick observatory was also in Russia for the eclipse, but

---

[109] Hasselberg's account of the expedition, *KVAÅ* 1915, 99-125, 122f.

[110] Such disappointment is documented among other eclipse observers. Pang, "Social Event", 274.

[111] They were later analysed by Jöran Ramberg. Jöran Ramberg, "Ljusfördelningen i solkoronan: Resultat från de totala solförmörkelserna 1945 och 1914", *PAT* vol 32 (1951), 100-120.

[112] Hasselberg's account of the expedition, *KVAÅ* 1915, 99-125, 122f.

was hindered both by clouds and political obstacles.[113] A request was made by W.W. Campbell for copies of Hasselberg's plates for inclusion in the Lick collection of large-scale corona plates, but such copies does not seem to have reached Campbell.[114]

Science and Photography in Sweden

Photographic technologies was a field where astronomers could communicate with scientists from other disciplines, as well as parts of culture outside of science. It is an example of a phenomenon not uncommon in the history of science: the frequent connections between scientists and society established through a scientific technology. Several astronomers participated in a wider photographic community in Sweden. Together with scientists from other fields, astronomers were present in the photographic culture that flourished during the second half of the nineteenth century. In fact, science in Sweden had taken an interest in photography early on. The Royal Academy of Science at Stockholm reported on the first photographic activities and early on – in the spring of 1842 – the chemist Jöns Jacob Berzelius and Fabian Wrede were experimenting with photography, together with the Academy's instrument maker. The departments of physics at Lund and Uppsala bought photographic equipment.[115] Later on, Carl Curman, an enthusiastic and skilled amateur photographer, started a department for photographic work in medicine at the Karolinska institute of medicine in 1861.[116] The meteorologist Hugo Hildebrandsson made a photographic study on the classification of clouds published in 1877, and the polar expeditions used photography increasingly.[117]

---

[113] John A. Eddy, "The Schaeberle 40-ft Eclipse Camera of Lick Observatory", *JHA* vol 2 (1971), 1-22, 15f.

[114] Campbell to Hasselberg 28 October 1918, Hasselberg's papers, 466i:3, UUB. There are no indications in the Hasselberg papers at Uppsala that I know of that the plates reached Lick, nor in the Shane archives at Lick observatory. Donald Osterbrock, private communication, October 1994.

[115] Rolf Söderberg & Pär Rittsel, *Den svenska fotografins historia* (Stockholm, 1983), 20f.

[116] Solveig Jülich, "Medicinen och fotografiets mekaniska objektivitet: Carl Curman och tillkomsten av Karolinska institutets fotografiska ateljé 1861", *Lychnos* 1998, 75-120.

[117] Söderberg & Rittsel, *Fotografins historia*, 64, 114f.

The 1880's saw increased photographic activity in Sweden. Photographic societies were formed: in Stockholm and Gothenburg in 1888, Uppsala 1889 and Lund 1893. Many of their members were scientists. In the societies professional photographers met with amateurs interested in photography, as well as scientists interested in learning more of the technology.[118] The Uppsala photographic society had as its goal to advance the use of photography as a tool in science, the arts and the industry.[119]

The scientists were important for the photographic societies. Scientists had in their scientific practice dealt with several issues of the photographic technologies that were of interest to other photographers. One example of this is Nils Dunér's systematic study of the relative merits of 124 recipes for developers published in the third and fourth volumes of *Fotografisk tidskrift*. The photographic technology was something that bound astronomy to society; photography was a technology that largely was developed outside of science, and then imported into astronomy. But there was also a flow in the other direction, from the scientific field to a wider field defined by the photographic societies. Astronomers were highly regarded members of the photographic culture, participating and giving their knowledge of photographic processes that they had developed for scientific reasons back to the photographic culture. The Photographic Society at Stockholm had the astronomer Hugo Gyldén as chairman; John Lundgren, professor of physiology and veterinary sciences was also on the board. The Uppsala Photography Society had Nils Dunér as chairman and professor of physiology Frithiof Holmgren as second chairman; among the members were scientists like the astronomer C.V.L. Charlier, professor of medicine Salomon Henschen, the physicist Robert Thalén and the physiologist Hjalmar Öhrvall. Photography was a trading zone, where astronomers and other scientists, even portrait photographers and amateur photographers, could gather around the practice of photography. Scientific technologies and methods of working are a very important reason for communication between scientists and society, between scientists from different disciplines and between scientists from different subdisciplines within a discipline like astronomy. Such a field of communication and articulation of standards of science – a reason for articulating standards and a vector for the transmission of various ideas

---

[118] Data for the societies and their membership is from *Fotografisk tidskrifts årsbok* 1894-1895 (Stockholm, 1895) 80ff.

[119] Statutes of the Uppsala photographic society, §1, Arvid Odencrantz papers UUB okat. 520 F.

throughout science and society - can be found by the historian when looking at the use of scientific technologies like photography.[120]

The photographic societies and the astronomers and other scientists that were active in them shows examples of the extensive contacts between scientists and other parts of society when it comes to a technology. Photography, like some other technologies used by the scientist, is situated between science and society, and serves as an important interface between science and society.

Photography, Mechanisation of Vision and the Objectivity of Observational Astronomy

Image technologies can be seen as tied to ideals of scientific objectivity. Mechanically produced images such as astronomical photography and x-ray imaging were typical in this trend which argued that nature should speak for itself. This trend of mechanised visual culture was developed against an earlier tradition where the scientist was invited to interpret and idealise to get at the truth behind the individual specimens. Rather than having a scientist interpret nature to find the typical from the individual specimen of a flower or a human being, a scientific ideal developed from about the 1830's that associated objective data with mechanical image procedures: nature should be automatically registered, and the scientist should not interfere too much; the scientist should take a step back, as it were, and also become more mechanic in his way of doing observations.[121] This tradition of mechanised objectivity is also present in the way astronomers thought about photography. The photographic astronomers were enthusiastic about the new technology, hailing it as one of the most important advances in observational astronomy ever made. For them, it signalled modernity. Especially when presenting themselves to the public, in popular astronomy, this enthusiasm is apparent. With the photographic technology astronomers could reach the very faintest of stars,

[120] Perhaps the instruments and other parts of scientific technologies is one important ingredient of the glue that holds together a fragmented discipline. Cf Galison, *Image and Logic*, 46-63.

[121] Lorraine Daston & Peter Galison, "The Image of Objectivity", *Representations* vol 40 (1992), 81-128.

"where the human eye is powerless".[122] Photography was regarded as more objective and reliable than the visual methods: after the "replacement of the human eye, which is always more or less subjective, with a retina that 'never forgets', the observations must reach a level of objectivity never before achieved".[123] The technology shielded the astronomer from the problematic raw nature of the direct visual observations in the darkened dome of the observatory. The stars were translated into clumps of silver grains on the exposed and developed plate and transported to the safety of the daylight. By studying the plate in the controlled conditions of the daytime observatory workplace, elements of error were screened out. This controlling effect of the photographic plate, this moving of stellar images from nocturnal nature to daytime desk in the office, was seen as one of the main points of photography. Photography was, in the opinion of Nils Dunér, reliable and objective, compared to visual methods, where the eye "became confused among the vast amount of shining points, and it became impossible to guard oneself against mistakes; confusion was almost inevitable. On a photographic plate, on the other hand, measurements could be made in the office, safe from such errors."[124] One theme in the rhetoric surrounding photography was that it made measurements more secure in that it allowed them to be performed in a more controlled environment. In the visual mode, the astronomer observed in the dark, it was cold, and often there was only one chance of doing an observation, such as in meridian observations; the nocturnal visual observations made mistakes common. With photographic observations of the sky, nature was framed, presented on a plate for repeated daytime analysis in a controlled environment.

Photography was seen as an enhancement of the human senses, and a technology that allowed astronomical observations to become controlled:

> The medieval alchemists dreamt of one day being able to produce an 'homunculus', a chemical product with all the properties of a man. Scientists of today are not as utopian; now they do not seek to make such a complex machinery as the homo sapiens, but they try instead to surpass the senses. The photographic plate is one of these and perhaps the most brilliant of them all, as

---

[122] Nils Dunér, "Om astronomisk fotografi", *Fotografisk tidskrifts årsbok* 1892 (Stockholm, 1892), 10-16, 14f.

[123] Bernhard Hasselberg, "Om totala solförmörkelser och deras betydelse för studiet af solens fysik, särskildt med hänsyn till den förestående solförmörkelsen den 21 augusti 1914", *Nordisk tidskrift* 1913, 21-40, 29.

[124] Nils Dunér, "Om astronomisk fotografi", 14f.

> it supplants vision, the most perfect of man's senses. [...] After an hour's exposure the stars have been transferred from the sky on to the photographic plate. Then the stars can be conveniently measured with a microscope[125]

The new technology even allowed astronomers in lesser countries to compete with the leading observatories, at least that was what Dunér claimed when discussing the new photographic equipment he was about to install at the Uppsala observatory: "with a refractor like the one that is proposed for the Uppsala observatory, it is possible to photograph stars that would be invisible even with the most powerful telescope in the world".[126] Photography was hailed as a more objective means of observing the sky, which also reached farther into space.

Photography was used in both classical and stellar astronomy and astrophysics. Photography could support astrophysical data collecting and classification. It could also be adapted to the classical fields of astrometry. Thus, astronomers could move from classical to modern astronomy; astronomers from the modern and classical fields could also gather around technologies like photography. The rhetoric of objectivity that surrounded photography could also have made the modern groups more accepted among the classicists. These factors could have been important in keeping astronomy a unified discipline, despite the emergence of the new fields. Photography found users both in astrophysics and classical astronomy; the practice of precision measurements on the photographic plates also gave the technology such a prominent place also with classicists.

A quantitative way of working was also tied to photographic astronomy. Astronomers did not simply 'take pictures' of the sky. After exposure and development, the plates often entered a quantification phase, where astronomers measured positions or densities of the plate. They spent large amounts of time on this stage. These measurements were made with plate measuring machines that provided astronomers with a way of repeatedly achieving a high precision of position measurement on the plates. Astronomers like Östen Bergstrand made much use of these ways of using photographic technologies. When Bergstrand measured the parallax of stars, or when Dunér measured the distance to the asteroid Eros, they did not 'look at pictures'; they measured the positions of the objects with very high precision relative to a number of standard stars on the plate.

---

[125] C.V.L. Charlier, "Om stjernfotografi", *Fotografisk tidskrift* vol 2 (1889) 251-254.

[126] Nils Dunér, *Några ord angående den föreslagna nya refraktorn å Upsala observatorium* (Uppsala, 1890), 4.

Photometric measurements were developed to analyse the brightness of celestial objects from the appearance on the plates. Stellar brightness data could come from a measurement of the diameters of stars on the plate; later, astronomers measured the density of the silver grains on the plate with a separate instrument, to gain knowledge of the brightness of the stars.

These steps were vital to the use of photography in astronomy and account for a way of doing astronomy that was very much a system of quantitative data and precision, an ideal that had old traditions in astronomy, since it was at the core of observational classical astronomy.

Charlier and Stellar Statistics

In June 1912 a group of young girls were picking flowers close to Vittsjö, in the south of Sweden. They were specifically looking for the "Skogsstjärna", Chickweed Wintergreen, *Trientalis europæa* L., that was growing there in large numbers. The girls, Carl Ludvig Charlier's daughters, collected 469 specimens during two weeks. To avoid a bias favouring conspicuous specimens, Charlier had instructed them to try to get all specimens in a defined area.

The first thing Charlier did when the daughters had collected the flowers was to establish a card catalogue, where he entered the characteristics of every flower. A card catalogue is, Charlier noted, "indispensable as soon as more than one character in a given population is considered."[127] The facts noted were the number of flower stalks, stamens, petals, sepals and leaves in the rosule. When all the facts had been measured and entered onto the index cards, a statistical survey of the plants could begin. Charlier thus got a collection of precise data on the *Trientalis Europæa*, numerical data with estimates of the dispersion around the mean, for the things that botanists usually described as 'common', 'rare', 'usually' and so on. Linneaus had observed that the number of stamens of the *Trientalis Europæa* varied, containing almost always more than five and most often numbering seven (therefore the plant became the type for the 7th class, Heptandria). With his data, Charlier could show that the probability that a specimen of *Trientalis europæa* L. had 7 stamens was 0.7514 ± 0.0329.[128]

[127] C.V.L. Charlier, "A Statistical Description of Trientalis Europæa", *Arkiv för botanik* vol 12 (1913) no 14, 2.

[128] Charlier, "Statistical Description", 4.

The work that Charlier performed at his summer residence during the summer of 1912 with ocular micrometer, index cards, and the assistance of his daughters collecting the flowers was not an attempt to make a breakthrough in botany. It was, as Charlier stated, a way for him to get some numerical illustrations to use in the elementary lectures on biological statistics he organised in Lund in the autumn of 1912. It might also be considered as an attempt to display his way of doing science to the botanists. Charlier noted that his method had been presented in his introductory text *Grunddragen af den matematiska statistiken*, (of which "[s]ome copies are still to be had", he also noted).[129] Though simple, the survey of the Trientalis Europæa contained several of the elements of Charlier's scientific style. There was a division of labour in the collection of data, bringing him material for analysis. There was also the organisation of a large amount of data in an index catalogue and the statistical treatment of empirical data. A general interest for this mode of work is shown by the fact that the results of the Charlier family's botanical excursion during their summer vacation resulted in a publication in the journal *Arkiv för botanik*, published by the Royal Academy of Sciences at Stockholm.

Charlier had begun this way of doing science some years earlier. Having left the classical field of celestial mechanics, he began to develop an interest in mathematical statistics and its application to astronomical problems. His first paper in this field dates from 1905.[130] He soon had published a number of papers on pure statistics, as well as an introductory text, *Grunddragen af den matematiska statistiken*.[131] While he had some plans early on to use the statistical methods on biological data, he soon began to pursue statistics as applied to stellar astronomy.[132] He started a large research programme for statistical analysis of stellar data, obtained at other observatories. Charlier organised a theoretical laboratory to which scientific data was brought from other institutions by way of mobilising technologies like the photographic plate and catalogues of stellar data. The material was used by staff working according to an organised and rational plan; out came models of the cosmos.

[129] Charlier, "Statistical Description", 2, n. 1.

[130] C.V.L. Charlier, "Über das Fehlergesetz", *Arkiv*, vol 2 (1905), no 8.

[131] C.V.L. Charlier, *Grunddragen af den matematiska statistiken* (Lund, 1910).

[132] In 1907 Charlier applied for funds from the foundation Lars Hiertas Minne to do statistical data on biological material, for example the anthropological data collected by Retzius and Fürst. Charlier to Stiftelsen Lars Hiertas Minne 28 February 1907, Astronomy department "Koncept, institutionshandlingar 1907-1928", page 2, LUB.

Carl Vilhelm Ludvig Charlier had a career that spanned celestial mechanics, observational photometry, stellar statistics, and cosmology. The main part of his work lay in the field of stellar statistics, a way of attacking the problem of the structure of the stellar system by using statistical analysis of large amounts of data on stellar position, brightness and motion.[133] Charlier's stellar statistics was coupled with the mobility of astronomical data produced by the photographic plate and catalogues of stellar data. A division of labour is present in his arrangement of the astronomers' work at the Lund observatory. The industrialisation of the astronomical workplace is evident in how Charlier organised work at the Lund observatory.

His work in stellar statistics was an example of the traffic in the interface between science and society.[134] Charlier's work in stellar statistics entailed a working-out of statistical methods as such. The knowledge of these statistical methods was a mobile competence, which could be transported to other parts of society by Charlier and his pupils. Charlier wanted to take part in the rationalisation of society, and his scientific work in stellar statistics became a means to that end.

The Young and Somewhat Radical Astronomer

Carl Vilhelm Ludvig Charlier (1862-1934) began studying at Uppsala in 1881.[135] In 1884 he had become assistant astronomer at the Uppsala observatory. From the beginning of his scientific career, Charlier concentrated on what was then called theoretical astronomy, that is celestial mechanics. In his PhD thesis, Charlier studied the perturbations caused by Jupiter on the asteroid Thetis. Hugo Gyldén at the Stockholm observatory was the leader in this field in Sweden, and Charlier used Gyldén's meth-

---

[133] The best introduction to the history of stellar statistics is Erich Robert Paul, *Statistical Cosmology*.

[134] This is one point where this treatment differs from Paul's. Here the development of stellar statistics is seen more in a context of mechanical calculators, unskilled labour, the use for statistics in government etc.

[135] Biographical material in Knut Lundmark, "Charlier, Carl Vilhelm Ludvig", *SBL*, and Gunnar Malmquist, "Carl Vilhelm Ludvig Charlier", *KVAÅ* vol 58 (1960), 385-405. Some details from Charlier's family are found in Essie Jansson, *Till minne av Siri* (Stockholm, 1988) and Essie Jansson, *Lundaminnen av studentska 1917* (Stockholm, 1989).

ods of calculating perturbations in the thesis. The work got good grades, and Charlier became *docent* on the thesis.

During his time as assistant astronomer he also worked in practical astronomy. He worked on stellar photometry with the Zöllner photometer at the Uppsala observatory, and observed comets.[136] After the thesis Charlier made a study trip to the observatories at Pulkovo and Berlin. After arriving back in Sweden, he got an appointment as assistant astronomer in Stockholm. Here he worked with the photographic determination of stellar magnitudes.

After a period as assistant astronomer in Stockholm he got a place as *observator* (assistant professor) at Uppsala in 1890. He took part in the routine work at the observatory, like observing the sun for determining the precise time of the equinox.[137] He also became known for his activities outside of the academy. Charlier had a radical view of society, though he was not one of the most radical. Earlier, during his student days, he had joined the radical student organisation Verdandi. He was also active in the popularisation of astronomy. The University of Uppsala started in the 1890's to arrange courses during the summer on popular science; the audience was mainly teachers, but also interested people from other walks of life were present. Here, the scientists had a chance to discuss their work and also the way they perceived science. Charlier gave a talk on the history of astronomy. This type of courses were soon taken up at the Lund University.[138]

---

[136] Uppsala observatory yearly report, *UUÅ* 1884, 42; 1885, 42; 1886, 64. Measuring the brightness of stars with the Zöllner photometer – constructed by the astrophysicist Zöllner in the 1860s – was done by comparing a star with an artificial star produced by the photometer. The latter could be artificially dimmed; the point where the two stars had the same brightness gave a measurement of the star's brightness. John B. Hearnshaw, *The Measurement of Starlight: Two Centuries of Astronomical Photometry* (Cambridge, 1996), 61ff. C.V.L. Charlier, *Astrophotometrische Studien* (*Bihang till k. svenska vetenskapsakademiens handlingar* vol 14 part I no 2) (Stockholm, 1888); Klaus Staubermann, *Controlling Vision*.

[137] Uppsala observatory yearly report, *UUÅ* 1891, 60; 1892, 62; 1893, 73; 1894; 75.

[138] In 1895 the summer courses had 467 participants (257 female, 210 male), 363 of which were teachers. *Sommarkurserna i Uppsala 1895: Förteckning öfver deltagarne* (Uppsala, 1895); *Sommarkurserna i Uppsala 1895: Tillägg och rättelser till förteckning öfver deltagarne* (Uppsala, 1895). Göran Blomqvist, *Elfenbenstorn eller statsskepp? Stat, universitet och akademisk frihet i vardag och vision från Agardh till Schück* (Bibliotheca historica lundensis, 71) (Lund, 1992), 287. See also "Sommarkurserna i Uppsala", *Svensk läraretidning* vol 12 no 35; J.A. Lundell, "De akademiska sommarkurserna i Uppsala", *Verdandi* 1894.

Charlier also lectured in other organisations. In the spring term of 1894 he lectured at the "Workers' institute" at Gävle, and in the next year he presented a talk in the society "Students and workers", formed in 1886 with the purpose of bringing workers and students together in discussions etc.[139] The talk was on the subject of religion, and it gives some insight in the kind of ideas Charlier had on matters outside of astronomy.[140] He argued that religions had to become more rational, they had to adapt to the findings of modern science. "Remove the supernatural, the superstition, the untrue from the ruling religions".[141] Religions had historically provided positive achievements in ethics, but these kinds of questions could now, in his age, be better dealt with "on the scientific basis of national economics or philosophical ethics".[142] Society was to be ruled by the truly rational and scientific insights, not by religion.

A main error in the big religions, Buddhism excepted, was that they all claimed to have got their texts directly from God; thus, intolerance was common when different cultures and religions met. Charlier put forward a notion of theological convergence, arguing that every religion had its own part of truth.[143]

Charlier's ideas on religions also came with a social perspective. Religions had traditionally tried to make it easier for poor people to endure the hardships of everyday survival by promising them justice in an eternal afterlife.

> Shall we feed the poor with fairy tales instead of bread, shall they be promised an imaginary paradise in a distant future that they might not reach, so that their slavery shall not be too hard to bear? No, we cannot be that heartless [...] we shall give him bodily food to eat and spiritual food that is healthy and nutritious and not founded on superstition. Let us try to solve the social issue, let us get hunger out of the world, so that it only remains as an unused word, appearing in quite horrible stories.

It is not difficult to imagine that with such ideas, Charlier came to be regarded as something of a radical person. That is one explanation for the debate that surfaced when he applied for the professorship of astronomy

---

[139] Crister Skoglund, *Vita mössor under röda fanor: Vänsterstudenter, kulturradikalism och bildningsideal i Sverige 1880-1940* (Stockholm, 1991), 86ff.

[140] C.V.L. Charlier, *Kan religion finnas utan gudomlig uppenbarelse?* (Uppsala, 1895).

[141] Charlier, *Religion*, 6.

[142] Charlier, *Religion*, 12.

[143] Charlier, *Religion*, 20f.

at Lund university in 1895. In this debate, conservatives were critical of Charlier's radicalism, but in the end, Charlier got the professorship.

The Theoretical Laboratory

After he became professor in Lund in 1897 Charlier continued to work on celestial mechanics for some years. He dealt with these problems in depth in his advanced textbook *Die Mechanik des Himmels*, issued in two volumes 1902-1907. The left-wing radicalism of his pre-professor days was rather mild; he never became what his conservative critics had feared. He was one of the founders of a leftist academic society, together with the philosopher Hans Larsson, the jurist Östen Undén, the astronomer and statistician Sven Wicksell, the philosopher Alf Ahlberg.[144] He was asked to submit articles to the liberal magazine *Forum*, but he declined.[145] Charlier never became that dangerous person to Lund University that his critics feared. Instead, he became a successful astronomer and scientific organiser.

*

Charlier's choice of field can be discussed from several angles. One important aspect is the national scientific context. There was a specialisation of work going on at the Swedish observatories. At the observatory in Uppsala the activities since Dunér's appointment as professor there in 1888 were mainly focused on observational astronomy, with modern equipment. To do something comparable, Charlier would have to convince the university or the government that his observatory should get substantial extra funding, which was not at all likely; politicians might very well have argued that one expensive and modern observatory was enough in Sweden. Uppsala's large advance in instruments is thus one factor to take into account. It is not even certain that Charlier wanted to compete with Uppsala in observational astronomy, maybe he wanted to do something different than the people at Uppsala.

Another aspect is the international context, where stellar statistics had been a rather large field since the turn of the century. Jacobus Cornelius

[144] Thorild Dahlgren, "Anteckningar från studentår 1907-1918", *Under Lundagårds kronor: tredje samlingen* (Lund, 1955), 231-308, 271.

[145] Torgny Segerstedt to Charlier 27 july 1914, Charlier collection, LUB.

Kapteyn in Holland and Hugo von Seeliger in Germany had started leading institutes working in stellar statistics. Kapteyn's Astronomical Laboratory at Groningen was a leading institute in the modern rational way of handling large amounts of stellar data.[146]

The stellarstatistical programme, which became Charlier's overarching field of work for the rest of his career, had connections with changes in scientific technology, organisation and institution planning. The very nature of being an astronomer was changing. Through the introduction of photography the speed of dataproduction at astronomical observatories had increased. Spectroscopy had given the astronomer new kinds of data. A whole new set of ways to cope with astronomical data had to be devised. To make it work, other ways of organising work were needed than the ones previously employed; the astronomer did not work any longer all by himself, or with small numbers of student assistants. To analyse large amounts of data, many people had to get involved in the process, and the astronomer had to device ways to make large groups of people – many of whom were not astronomers – work effectively together.

Activities at the Lund observatory were organised according to Charlier's rational plans. He wanted to create a modern institute of scientific number-crunching, analysis, and thought – not an institute organised on old-fashioned principles, like the one he had seen when he visited the Nautical Almanac Office in London. (Charlier was more positive towards the Greenwich observatory.)[147] The division of labour that was the talk of the times also reached astronomy. There was an older division of labour in that the theoretical astronomer worked with theory and other astronomer's empirical data, and the observational astronomer observed the sky. To this had been added the possibility, through the introduction of the photographic plate, of working with large sets of previously unanalysed data without observing the sky. There was also a division of labour in lesser scale, in the everyday activities at the individual observatory. Charlier's ideal mode of work lay far away from the often lonely work of the classical astronomer; other times had arrived, it was even less necessary than before for the individual astronomer to participate in every step in the process. With a corporate metaphor, one could call Charlier's mode of work 'industrial', where every single individual did a

[146] For the history of stellar statistics, see Robert Paul, *Statistical Cosmology*.

[147] Charlier's travel report dated October 1907, Astronomy department "Koncept, institutionshandlingar 1907-1928", page 6, LUB.

clearly defined part of the total process. It differed from the older astronomy, in which the 'craftsman' excited total control over large pieces of the production process. In this aspect, Charlier's way of running a theoretical laboratory mirrors the industrialisation of American observational astronomy.[148]

Another aspect is the cultural context, where changes were taking place in the workplace and workforce; changes in the mode of operation of an astronomical observatory do not necessarily come from within science. To further his project, Charlier connected with allies from all these contexts. He tried to tie together these heterogeneous cultural, technological and scientific resources in his institute. The possibilities presented by changes in society were seized upon by Charlier, who tried to draw them close to the observatory, besides trying to get other astronomical institutes aligned with his institute. He wanted to establish a laboratory for theoretical astronomy, and construct a quite complex mosaic of astronomers, computers, calculating machinery for working on large amounts of astronomical data; this was to be his instrument at large for doing theoretical astronomy. *Computers* in those days meant people doing number crunching: they were an important part of Charlier's vision of the modern institute.

In his vision of the modern institute, astronomers ought to be removed from the time-consuming chore of making large amounts of quite simple calculations. A statistical treatment of data on stars entailed large amounts of numbers being moved about in routine calculations that often could be reduced to almost mechanical operations. The astronomer ought to get rid of this and instead be able to concentrate on larger issues, such as strategic thinking. Computers were necessary to the modern astronomical institute:

> Regardless of whether an astronomer's work consists of theoretical calculations of the motions of heavenly bodies or in the determination of their places or physical nature through observations, always or nearly always a quantitatively large part of the necessary work is calculation of a very elementary nature, which every person with an ordinary school education easily can learn to perform. If this elementary calculation must be done by the astronomer himself, his work will progress very slowly and his interest in executing such studies must become to a large degree diminished. Therefore, it is an inexcusable

[148] John Lankford and Ricky L. Slavings, "The Industrialization of American Astronomy, 1880-1940", *Physics Today*, January 1996, 34-40.

waste of energy, if such computers of elementary schooling are not put to the astronomer's disposal.[149]

That was how Charlier argued when pleading for 1500 kr of extra funding for computers in 1907. People doing routine computing had of course existed before in astronomy, but then they had often been recruited from other kinds of personnel. Students or astronomers in the early stages of their career working extra were earlier the typical computer. Now female computers without any training in astronomy or science were introduced into Swedish astronomy by Charlier. Women had entered the mechanised worlds of offices and communication systems – the telephone switchboards, the typing bureaux, the telegraph stations – and in 1907 that category of personnel reached Lund observatory. They became an important part of Charlier's workforce.[150] The existence of a workforce that was inexpensive to hire and had an adequate schooling – many of the computers at the observatory had got secondary schooling but they were not astronomers – meant much for the industrialisation of the work at the Lund observatory.[151]

The women at the observatory can be seen as an effect of change in the labour market. Women appeared at workplaces, as well as in science. In a study of laboratory assistants, Bodil Persson has pointed to the large numbers of assistants doing much work. Salomon Henschen was an early adopter, hiring seven women assistants for his brain research programme in Uppsala during the 1880's and 1890's. They made thousands of preparations, thin slices of brain tissue used by Henschen as he analysed his way through the brain; they also made the drawings that were an important part of Henschen's publications.[152] Persson thus lifts forward a previ-

---

[149] Yearly report of the astronomy department 1906-1907, *LUÅ* 1907, 24-26, 26.

[150] An introduction to women in the public bureaucracy is Christina Florin, "Statsbyråkratins erogena zoner: Staten som mötesplats för kvinnligt och manligt", Gunnar Broberg, Ulla Wikander, Klas Åmark eds., *Svensk historia underifrån: Bryta, bygga, bo* (Stockholm, 1994), 174-198. Women in the communications industry: Torsten Althin, *Telelandet Sverige* (Stockholm, 1953), 48-56.

[151] Walter Gyllenberg claims that every computer present in 1927 had studied at secondary schools for women. Gyllenberg in a report to the academic senate 21 April 1927, Astronomy department "Koncept, institutionshandlingar 1907-1928", 523f, LUB.

[152] Bodil Persson, *När kvinnorna kom in i männens värld: Framväxten av ett kvinnligt yrke – Laboratorieassistenter under perioden 1880-1941* (Stockholm, 1994), 105-190. Cf. Torbjörn Gustafsson, *Själens biologi: Medicinen, kulturen och naturens ordning 1850-1920* (Stockholm & Stehag, 1996), 79-137.

ously rather unknown workforce in the history of science in Sweden. Jenny Beckman has discussed Axel and Thérèse Ekblom who, with their children, all worked in Stockholm at the National Museum of Natural History and the Bergian botanical garden as illustrators, assistants and secretaries. Lund observatory was not the only place where assistants, often female, were doing routine work. These otherwise invisible technicians and assistants deserve more studies from historians of science.[153]

The entrance of computing assistants was also seen in connection with a change in office technology. Calculating machinery made the routine calculations easier to standardise. For Charlier, the existence of calculating machinery was vital for the entrance of the new workforce. "Fortunately the recent advances in practical calculating machinery makes the use of relatively cheap employees possible - women [*fruntimmer*] with ordinary schooling and a talent for numerical calculation can do a large part of the necessary work", Charlier wrote in a petition for funding in 1908.[154] Scientific practice at the observatory was in several ways connected to what happened in the working life in the surrounding society, both as regards the availability of a suitable workforce and suitable office technology. The activities at the observatory were changed because of the modernisation of the living space, begun towards the end of the 1800's, to use a term from Anders Ekström. What used to be a rather untechnical office workspace, with pen and paper, was mechanised.[155] Charlier saw how this change in society at large could be used for his project to modernise astronomy at the Lund observatory. The calculating machinery and the women who worked in the increasingly mechanical office workspace of the day were resources that Charlier drew upon to advance his scientific work. A skilful leader of a scientific institute was able to manage such a drawing together of heterogeneous resources and practices. In astronomy, such management skills were in the days of Charlier becoming important, as the practice of astronomy developed its complexity and size. The successful professor and leader of an observatory was no longer defined only by the brilliance of his own

[153] Jenny Beckman, "Vetenskap och konst: Illustratörsfamiljen Ekblom vid Naturhistoriska riksmuseet", Marika Hedin & Ulf Larsson eds., *Teknikens landskap: En teknikhistorisk antologi tillägnad Svante Lindqvist* (Stockholm, 1998).

[154] Argument for further funding, Astronomy department "Koncept, institutionshandlingar 1907-1928", 9, LUB.

[155] Anders Ekström, *Den utställda världen: Stockholmsutställningen 1897 och 1800-talets världsutställningar* (Nordiska muséets handlingar 119) (Stockholm, 1994), chapter 2.

thoughts and observations; the successful leader of an observatory had to be a manager, an entrepreneur, an organiser.

The observatory acquired several mechanical calculating machines, one of which was a motorised machine, delivered by the Swiss engineering firm of Hans W. Egli.[156]

Work at the Lund observatory revolved around the calculating office [*räknekammare*] where the computers used the calculating machinery, or worked with paper and pen. At the Lund observatory, the calculating office supplanted the dome as the primary place of astronomical work. Lund observatory under Charlier was characterised by a rather strict discipline that came with the way Charlier organised the work at the observatory. When Axel Corlin arrived in Lund as amanuensis in 1921, he gave a first impression of the work in Lund, as compared to what he had experienced at the observatory at Uppsala.[157] Corlin had to spend three hours each day doing calculations on Charlier's stellar statistical project. He felt disciplined. In the large calculating room five women were busy calculating and writing. One could talk with each other, but not too much – the discipline, Corlin thought, was just like in any office outside of science. Charlier, known as "the boss", was walking in and out of his room all the time, to get papers or control the work. Charlier sometimes could get quite uncontrollably angry, if the order of papers and numbers was not the one he demanded. Corlin complained about Charlier's personality, who was not at all as easy to get along with as his former professor at Uppsala, Östen Bergstrand. As amanuens, Corlin had to see to it that the observatory clock at the entrance was correct to the second; he also feared the day when he had to start teaching introductory courses. This letter was written when Corlin had been in Lund a short time, and in some subsequent letters the tone is a bit more positive, but still it gives an insight into how Charlier ran the place and how the new way of doing theoretical astronomy worked. Corlin was afraid of losing his identity as a free student and instead begin to feel like an office clerk. After a year spent at the Lund observatory, Corlin still longed for Uppsala – he felt isolated in Lund and did not particularly like the spirit and company at the Lund observatory.[158]

---

156 Charlier to Hans W. Egli 10 February, 7 March and 4 June 1912, Astronomy department "Koncept, institutionshandlingar 1907-1928", 35, 37, 40, 55, LUB.

157 Axel Corlin to Knut Lundmark January 17 1921, Lundmark's papers, LUB.

158 Östen Bergstrand to Knut Lundmark December 20 1921, Lundmark's papers, LUB.

The first female computer at the Lund observatory (and in Swedish astronomy) was Elin Bruzelius, who was employed in 1907, when she started working on asteroid orbits. The observatory had not at first got the extra funding mentioned above; Bruzelius was hired on money in the ordinary budget - an indication of the importance Charlier attached to the addition of computers to the staff. Bruzelius then went on to work at the observatory for a long time.

The computers handling the numbers where directly involved with the scientific work - the numbers they handled represented data on the brightness, motion, distance of stars; the very stuff that makes up a cosmology. Compared to women working at the same time at American institutes, like Harvard observatory, the Lund computers were far away from the scientific analysis. Harvard women got to work on more analytical problems. Like the Lund women, they had an energetic scientific entrepreneur, administrator and finance gatherer above them (Edward Pickering). They too had to work on routine tasks, like writing and calculation. But several of them worked on the classification of spectra; and both Antonia Maury and Annie Jump Cannon developed classification systems for spectra. Cannon also had planning responsibility for the large programs of spectral classification and she handled the publication of the important Henry Draper Catalogue after Pickering's death (besides, of course, doing the classifications).

The American women in astronomy had often studied science and astronomy at higher levels, many of them were PhDs. The American women had a very rough time trying to piece together a life in science with being women, but they had at least studied at a higher level, and several of them got to work on more advanced topics than just routine work. In Sweden, women did not study astronomy. Sweden did not have institutes like Vassar College, Radcliffe and so on. In his study of American astronomy, John Lankford's picture of the possibilities of women in American astronomy is a rather negative one. The Swedish case was even less positive; in Sweden women were not even studying astronomy.

In Sweden, the first woman to reach a higher education was Frida Palmér, who in 1934 got a fil.lic. degree; some years afterwards, she finished her PhD thesis and moved on to become a teacher at a secondary school in Halmstad.[159]

[159] The activities at Harvard: Hearnshaw, *Analysis*, chapter five. Palmér: Knut Lundmark, *Astronomien i Lund 1667-1936*, 72. Cf. John Lankford, *American Astronomy*, chapter nine. Cf Eva Isaksson, "Inte en himmelsk syn: kvinnor i Helsingfors

The women at the Lund observatory never developed any scientific classification system, published no works, got no positions in middle management, and they had not even studied astronomy. In Lund the hierarchy was also very strict. The computers were treated like secretaries, for instance, would have been treated at any other workplace.[160] Still, they were not exchangeable; one assistant could not always take another one's task. They developed skills that were valuable and in some cases unique, and it happened that astronomers and their tasks were connected with specific assistants. Take for example the case of reduction of meridian circle observations, the domain of Walter Gyllenberg. When the newly appointed successor to Charlier, Knut Lundmark, in 1929 "used miss Jansson for typewriting, Gyllenberg was very nervous. She is, you see, the only one of the women here capable of reducing meridian circle observations, and he needed her all the time [...] instead, let miss Lindström do the typewriting."[161]

Society was being modernised. Rational principles and analysis was to be the basis for much. Scientific management, with its main principle "every work process should be studied scientifically and in great detail in order to arrive at the best way to implement it" was thought to be the solution to many problems.[162] In administration and in the office space machines made their entry. The handling of words and numbers was mechanised. Where pencil and paper had been in use earlier, typewriters with their homogeneous writing style and ability to make copies came instead; the tables, pens and papers of calculations were exchanged for the calculating machines. This is a context in which one could see Charlier's theoretical laboratory, just as a traditional laboratory/observatory can (and should) be seen in a context of engineers, instrument makers, machinery, disciplined workforces, and the general level of technology in the surrounding society. Laboratories are in some ways products of a wider scientific, industrial, economical and military culture; the cultural

---

Carte du Ciel-projekt, 1893-1930", in Elisabeth Hermodsson & Lena Trojer eds., *Ikaros med två bröst: kvinnan och naturvetenskapen* (Stockholm, 1995), 79-85.

160 See for example the tone in a letter from Elin Bruzelius to Charlier 8 March 1924. On Elin Bruzelius and Charlier's daughters (who used to play at "Zelius" workplace at the observatory, see Jansson, *Minne*, 62f.

161 Axel Corlin to Knut Lundmark 28 April 1929, Lundmark's collection, LUB.

162 Quote from Hugo Avalder, "Arbetets vetenskap", *Nordisk tidskrift i organisation* vol 1 (1919), no 9, 118. This journal published many articles on scientific management. Cf Nils Runeby, "Americanism, Taylorism and Social Integration", *Scandinavian Journal of History* vol 3 (1978), 21-46.

embeddedness of science is evident in the many links that bind the laboratory/observatory with the surrounding society.[163] These links can be said to stabilise the observatory. Without these numerous links, an observatory/laboratory would lack the ability to produce scientific data. Thus, the observatory director's ability to stabilise his observatory by stringing the institute to resources was an important activity. The scientific field that dealt with calculations became modernised along these lines during the early twentieth century.[164]

The stabilisation of the observatory by the drawing together of resources to the laboratory is mirrored by a flow in the other direction. A process of interactive stabilisation is evident in Charlier's embedded laboratory/observatory. As will be shown later in this chapter, a number of statistical applications flowed from Charlier's theoretical laboratory to the society at large.

Another important factor was the international situation in astronomy. In America, astronomy had since the 1880's changed in a direction that John Lankford and Ricky Slavings have called "industrial": vertical organisational structures, large, well-planned projects, division of labour that made people with little amounts of education or professional status do data-processing, mechanised data acquisition and large amounts of money. George Ellery Hale, Edward C. Pickering, Simon Newcomb, Lewis Boss were some American astronomer-entrepreneurs who had figured out how to do modern astronomy on a large scale.[165] When Charlier got his own version of this kind of astronomy going, he dedicated the first part of his series *Studies in Stellar Statistics* to Edward Pickering.

Closer to Charlier, Kapteyn's work at Groningen was another example. He had worked out a way of doing observational astronomy without

---

[163] Cf. Svante Lindqvist, "A Wagnerian Theme in the History of Science: Scientific Glassblowing and the Role of Instrumentation", in Tore Frängsmyr ed., *Solomon's House Revisited: The Organization and Institutionalization of Science* (Canton, MA, 1990), 160-183; Peter Galison, *Image and Logic*, 8, 55.

[164] Andrew Warwick, "The Laboratory of Theory or What's Exact about the Exact Sciences?", M.Norton Wise ed., *The Values of Precision* (Princeton, 1995), 342. Warwick shows in this paper an interesting way of looking at theoretical sciences around the turn of the century. Warwick studies calculations and the way calculations are performed, a field that can be placed between experimental and theoretical science – a space Warwick calls *the theoretical laboratory*.

[165] John Lankford and Ricky L. Slavings, "The Industrialization".

telescopes, using photographs and other kinds of data transported to his Astronomical Laboratory from other observatories.[166]

Building, Unit, Seismograph and Clock: Signs and Science

The new mode of doing astronomy was also manifested in the buildings of the observatory.[167] When Charlier had arrived at Lund, not much space was left for expanding activities. A new and larger staff entailed a need for larger buildings. In several years Charlier applied for funds for a new building and when the money finally came, he constructed a building that did not look like any other in Swedish astronomy. It is impossible to tell just from the outside that this building is an astronomical observatory; it might just as well be an office building. In a way, the building – with its lack of towers or domes with telescopes – signals what kind of activity the Lund observatory was working on.

Charlier also tried to expand the activities at the observatory and enlarge the activities of his observatory across borders of the disciplinary mosaic that existed at the Lund university. In 1906, 7Charlier argued that a seismograph should be added to the instruments at Lund. Knowing the interior of the earth would help us know more about other planets was his argument for seismology as being part of astronomy. When the matter was dealt with in the mathematic-scientific section of the philosophical faculty at Lund in 1907, the proposal was not accepted. A.V. Bäcklund, professor of mechanics and mathematical physics, argued that seismology belonged to meteorology, and H.H. von Schwerin, assistant professor of geography and history argued that it rather belonged to physical geography.[168] Charlier did not leave the question and in 1912 the parliament assigned funds for a department of seismology at the observatory.[169] At a visit to Göttingen in 1913 Charlier met with professor Johann Emil Wiechert at the institute for geophysics. He also met with the instrument maker firm of Georg Bartels, where he ordered a seismograph of Wiechert's construction. Wiechert promised his institute would ana-

---

[166] Erich Robert Paul, *Statistical Cosmology*, 80ff.

[167] See Hjördis Kristenson, *Vetenskapens byggnader* for a study on the architecture of laboratories and observatories.

[168] Schalén, Hansson and Leide, *Observatoriet*, 87f.

[169] Yearly report of the observatory, *LUÅ* 1913, 39-42, 42.

lyse the seismograms that would be produced at the Lund observatory.[170] Complications because of the war delayed the instrument, but it was finally delivered in the autumn of 1916; measurements began on the first day of 1917. A fight against various hardships started. The reduction of the measurements proved to be a continuous problem; the rather damp ground on which the observatory is constructed made the seismograph cellar prone to be filled with water (sometimes from the city's sewage system); increasing traffic from the nearby railroad disturbed measurements.[171]

Calibration of the seismograms required a precise time-keeping to allow comparison with measurements in other parts of the world; hence it was important that the observatory acquired a means of securing an accurate time, fitted to international standards.

Time signals of high precision were being transmitted through radio, and Charlier was keen to keep the Lund observatory on a par with the latest time-keeping technology.[172] A radio receiver was ordered from Germany. Because of the outbreak of the war, the receiver did not arrive; an assistant at the department of physics in Lund, Albin Nilsson, instead constructed a receiver that functioned well. Time signals from Nordeich in Germany and the Eiffel Tower were received.[173] The time signals were used to calibrate a clock, made by Siemens-Schuckert of Berlin.

Such precise time-keeping equipment at the observatory also had other uses besides an aid for seismology. Astronomy was a science associated with precision, a bedrock of precise numbers in a world undergoing change. The precise time was used to calibrate the seismograms, but Charlier also made rhetorical use of it, when it was used to drive a clock that was put on display at a prominent place: by the entrance gate to the observatory grounds. This clock was intended for the public. The clock

---

[170] Charlier's travel report, travel to Göttingen and Hamburg 23 July - 18 August 1913, Astronomy department "Koncept, institutionshandlingar 1907-1928", 70, LUB.

[171] The seismograph was dismantled in 1954. Interview with Nils Hansson by the author, 1993. Charlier's report to the chancellor for the Swedish universities 15 October 1913, Astronomy department "Koncept, institutionshandlingar 1907-1928", 68f LUB; Charlier's report 28 November 1914, Astronomy department "Koncept, institutionshandlingar 1907-1928", 108 LUB; Schalén, Hansson, Leide, *Observatoriet*, 92.

[172] During a conference in Paris in 1913, plans were made for the international standardization of time, distributed through radio time signals. Charlier participated in the conference as representative from Sweden. Charlier's report to the minister of culture Fridtjuv Berg 21 December 1913, copy in Charlier's collection, LUB.

[173] Yearly report of the Lund observatory, *LUÅ* 1914-1915, 39-43, 41.

face is a large one showing seconds, inside this is a lesser hand showing the hours; seconds are prominent. The clock was to signal precision to the citizens of Lund. "The main purpose with the clock I intend to display at the observatory is that the seconds should be made prominent – like the clock at Greenwich", Charlier wrote the maker of the clock in 1915.[174] The design was a success: the clock became one of the more visible parts of the observatory, and the citizens of Lund could calibrate their watches as they passed by the observatory gate with "the most accurate clock in Lund" as it was known – until someone smashed the clock a night in January 1974.[175] During the years the clock mechanism itself was repaired, changed, and eventually replaced by an electrical clock. Only the face open to the public was kept the same, which remained a visible sign of precision.

The radio distributed time signals had a double use at the observatory. They were vital for making use of the seismograph. They also had a rhetorical role to play. Scientific instruments sometimes play such multiple roles, pointing at various directions.

Charlier introduced his own unit for measuring distances in the cosmos. The cosmological models produced at the Lund observatory were all presented in the same unit of Charlier's invention: the siriometer. Charlier introduced the siriometer in 1911, when he published his first work in stellar statistics, *Constitution of the Milky Way*. Hugo von Seeliger had, before Charlier, introduced a similar unit, the siriusweite, defined as the distance to a star with a parallax of 0".2. Charlier's unit, at about the same size, is one million times the earth's distance from the sun (which translates to 15.79 light-years). He also introduced a new unit for time, the stellar year, defined as one million years.[176]

The light-year had been in use before, not only in popular works. Siriometer was used by Charlier and his staff in a majority of their works. When Gustaf Strömberg, who had spent some time with Charlier before going to the US, was about to publish in the *Astrophysical Journal* he tried using siriometer and stellar year, without success. He speculated

---

[174] Charlier to Siemens-Schuckert Stockholm 31 December 1915, Astronomy department "Koncept, institutionshandlingar 1907-1928", 140, LUB. Charlier's emphasis.

[175] The clock face was repaired for the cost of 1000 kr. "Observatorieuret sönderslaget", *Skånska Dagbladet* 29 January 1974; "Gammal klocka sönderslagen", *Arbetet* 29 January 1974; "Rättaste klockan i Lund nu lagad", *Arbetet* 11 January 1975; Lundmark, *Astronomien*, 15.

[176] C.V.L. Charlier, *Introduction to Stellar Statistics* (Lund, 1921), 7f.

that it might be because of courtesy towards Kapteyn. "It will probably not be long before the units are accepted" he reported optimistically back home to Charlier in 1918. He was wrong; the units never came to be used outside Lund astronomy, despite Charlier's and his pupils' use of them in their own publications and their struggles with editors at central journals.[177] "Perhaps I am too conservative in feeling that the light-year is best because of established usage, convenient size, and ease of comprehension" was the polite answer from Heber D. Curtis, editor of *Publications of the Astronomical Society of the Pacific*.[178] Siriometer never reached widespread use; instead, the International Astronomical Union in 1922 adopted the unit parsec.[179]

An International Institute

Charlier did not stop at reorganising the activities in Lund; he had plans for a larger international institute doing theoretical astronomy. One reason for him to discuss such plans was the asteroid problem. The introduction of photographic observations had made the number of discovered asteroids rise to very high levels. Theoretical astronomers had problems keeping pace with the discoveries. Observational astronomers discovered asteroids faster than the calculators could determine the orbits. A situation in which the discovered asteroids went astray, and disappeared out of sight before their orbits had been calculated was not far away. That would be a totally unacceptable state for a science whose self-image included notions of control, prediction, exactness.

In 1913 Charlier published his proposal to solving the problem as *A Plan for an Institute for Theoretical Astronomical Research*, in which he made a plea for theoretical astronomy. Observational astronomy had during several decades got larger resources because of both philanthropy and state funding. In this expansion of astronomy, theoretical astronomy seemed to have been forgotten. People had misunderstood the ways theoretical astronomers worked, thinking that brain, pencil and paper were sufficient for their work. This was far from the truth, Charlier reasoned,

---

[177] Gustaf Strömberg to Charlier 11 February 1918, Charlier's collection, LUB; copy of Charlier to Edwin Frost n.d. 1924, Charlier's collection, LUB.

[178] Heber D. Curtis to Charlier, 15 September 1913, Charlier's collection, LUB.

[179] K.G. Malmquist, "Några ord om avståndsenheterna inom stellarastronomien", *PAT* vol. 4 (1925), 46-48; Lisbeth Moustgaard, "Siriometer, siriusvidde og parsec. Enheder for afstande til stjernerne – den historiske udvikling", *AT* vol 24 (1995), 14-19.

and he proposed a radically different theoretical astronomy, based on the principles of division of labour: an astronomy taking place in a modern, almost office-like organisation and milieu, like the one he had planned and executed at the Lund observatory.

Charlier identified three stages in solving problems in theoretical astronomy. The first was that the astronomer should try to state the problem in a way that was solvable. When this abstract level had been attained, the next step was to make the systems of formulæ fit for numerical calculation. This second stage could be time-consuming but did not require original or deep mathematical skills, just an elementary mathematical education. The third level was the strictly numerical handling of the products from stage two, a level that demanded even less mathematical skills.[180]

The major problem for theoretical astronomy was that every single astronomer often had to execute all three levels himself, despite the fact that it only was the first that required a qualified astronomer. Lack of funding was the reason for this. To Charlier, advances in theoretical astronomy had been modest since the days of Lagrange and Laplace, because of this problem in scientific organisation. The solution was to organise a large international institute.

Charlier thought of an institute with three kinds of personnel, each doing a single level of the three-level model of how to solve a problem. The computation of asteroid orbits was to be only one part of the institute's work. He envisioned a first level consisting of eight astronomers, three of whom would be doing asteroid work, two studying the three body problem, two working on the shape of planets and tidal phenomena, and one astronomer doing stellar statistics. The wages should be higher than for the typical professorial chair to make sure the best astronomers would join the institute.

The eight astronomers should be supported by ten algebraic computers – level two – recruited from students that trained in the methods of theoretical astronomy during their temporary work at the institute. Numerical calculations – level three – could be done with mechanical calculators "and can hence, generally, be executed by lady-computors."[181] The trust in women doing calculations was grounded in them operating

---

[180] Anon. (=Charlier), *A Plan for an Institute for Theoretical Astronomical Research* (Lund, 1913), 3f.

[181] Charlier, *Plan*, 6.

machinery instead of calculating with pencil and paper. Calculating machinery gave authority to the results made by unskilled workers.

Cost for such an institute was estimated at an annual 200 000 German marks for expenses, one time expenses for buildings and for computing machines: 600 000 marks. The published work does not tell where Charlier thought the money would come from, but archival material reveals the source Charlier hoped could fund such an institute. There were two alternatives, an international co-operation with support from the governments of various states, or philanthropy "in which case Andrew Carnegie has been thought of as the only alternative." Charlier had approached the Swedish prime minister Lindman to try to figure out which Swede might be approached as a contact channel with Carnegie.[182]

In the winter of 1912, before the *Plan* was published, Charlier wrote Carnegie, enclosing a copy of his manuscript for the *Plan* with the letter. He proposed that Carnegie would study the plan, which, Charlier guaranteed, would become a worthy parallel to the great astronomical institutes Carnegie had funded in the US.[183] Charlier wanted to become the George Ellery Hale of theoretical astronomy. An answer from the Carnegie foundation to Charlier has not been found.

Other members of the Gesellschaft's committee on asteroids were critical. Rudolf Martin Brendel, professor of astronomy in Frankfurt aM had also plans for an institute with assistants; (with the difference that he wanted the institute to be placed in Frankfurt). They discussed how to get the French involved in the project.[184] Charlier's vision was too much for Brendel. Such a big institute would entail a centralisation of resources, leaving too much power to the eight astronomers of the institute. He also reacted at the very way of doing science and wanted to defend freedom for the individual researcher. Good scientific work, Brendel argued, could not come out of too much planned work, and could not be ordered forward by command. He would, however, support the proposal if it meant that the whole community of theoretical astronomy would gain from the

---

[182] Copy of a letter from Charlier to unknown, nd., Astronomy department "Koncept, institutionshandlingar 1907-1928", 21f, LUB; copy of a letter from Charlier to unknown, 15 June 1912, Astronomy department "Koncept, institutionshandlingar 1907-1928", 41, LUB.

[183] Copy of a letter Charlier to Carnegie 26 November 1912, Charlier's collection, LUB (placed among letters from C.R. Miller, editor with the New York Times, to Charlier).

[184] Brendel to Charlier 15 November and 25 November 1910, 26 May 1912, Charlier collection, LUB.

proposed institute.[185] Other members of the committee on asteroids – Fritz Cohn, Kobold and Witt – were slightly critical to the proposal, writing a joint report in which they aired their criticism.[186]

Charlier wanted with his proposal to make the production of theoretical astronomy more rational and effective, through the use of division of labour. He wanted to do on a larger scale what he had done in Lund. An astronomer calculating on his own was something that belonged to the past; the way forward was an increase in scale and a change towards a more collective and a more industrial way of working. It was a proposal to change what it meant to be a theoretical astronomer. It met with criticism; fellow astronomers did not want to give up their freedom, their individual control of the scientific work.

Perhaps the differences of opinion could have been solved, and Charlier would ultimately have been successful in obtaining funding. Instead, the first world war came, making international co-operation very difficult. Even during the 1920's the effects of the war was felt in astronomy. The war effectively stopped much of international co-operation and thus hindered Charlier's plans.

Charlier's Cosmos

The classical way of dealing with the advance of the perihelion of Mercury's orbit – one of the effects later explained by the theory of relativity – was to propose that there were unknown bodies in the inner solar system; one planet, several smaller asteroids or perhaps clouds of meteoritic matter. In 1912, Charlier argued that the most likely spot for such bodies ought to be the points where the bodies would form a equilateral triangle with the sun and Mercury, the Lagrange points. He stated his intention of looking for such objects at the coming solar eclipse in 1914.[187] Charlier gave the problem to Henrik Block, one of his students, who discussed whether the stability of such a point would be influenced by the strong eccentricity of Mercury's orbit.[188]

---

[185] Brendel to Charlier 27 June + 4 July 1912, Charlier's collection, LUB.

[186] Fritz Cohn to Charlier 6 July 1912, Charlier collection, LUB. At this place is also the undated report from Cohn, Kobold and Witt.

[187] C.V.L. Charlier, "Das Bodesche Gesetz und die sogenannten intramerkuriellen Planeten", *AN* no 4623 vol 193 (1912), 260-272.

[188] Henrik Block, "Sur la stabilité d'une petite planète aux points triangulaires de Mercure", *Arkiv* vol 10 no 4 (1914).

When the Lund observatory sent out an expedition to the eclipse of the sun in 1914, the main task was to look for such bodies; the results were negative. Later on Charlier continued to try to find such bodies. Writing to W.W. Campbell before the Lick observatory sent out a solar eclipse expedition in 1923, he asked if the expedition could take photographs to look for such intramercurial bodies. The expedition was unable to take the photographs Charlier had requested. Campbell proposed instead that the Lick observatory would try to photograph the area outside of eclipse. The areas he was interested in ought to be about 30 degrees from the sun, and therefore possible to photograph during the early evening or morning. Charlier sent him an ephemeris of the positions for the Lagrange points. Plates from previous eclipses were also studied in the Lick archive. No intramercurial bodies were found, that could explain the changing orbit of Mercury using classical mechanics.[189]

*

The main problem for stellar statistics was the distribution of stars in space. By using statistical methods on the large amounts of data collected on properties of stars, they intended to produce a picture of how the stars were distributed in space, and hence delineate the Milky Way system.

The first study made by Charlier in stellar statistics was published in 1912. In part it had an origin in lectures held for his graduate students. Choosing to concentrate his teaching duties to more advanced lectures, Charlier left more basic lecturing to other personnel at the institute.[190]

---

[189] Copies Charlier to Campbell 3 July 1923, 19 April 1923, Campbell to Charlier 3 August 1923, Charlier collection, LUB. Jeffrey Crelinstein, "William Wallace Campbell and the 'Einstein problem': an observational astronomer confronts the theory of relativity", *HSPS* vol 14 (1983), 1-91; John A. Eddy, "The Schaeberle 40-ft Eclipse Camera of Lick Observatory", *JHA* vol. 2 (1971), 1-22; Richard Baum & William Sheehan, *In Search of Planet Vulcan: The Ghost in Newton's Clockwork Universe* (New York & London, 1997), 247ff. For the reception of relativity in Sweden, Carl-Olow Stawström, "Relative Acceptance: The Introduction and Reception of Einstein's Theories in Sweden, 1905-1965", in Svante Lindqvist ed., *Center on the Periphery: Historical Aspects of 20th-Century Swedish Physics* (Canton, MA, 1993); Thord Silverbark, *Fysikens filosofi: Diskussioner om Einstein, relativitetsteorin och kvantfysiken i Sverige 1910-1970* (Stockholm/Stehag, 1999).

[190] C.V.L. Charlier, *Studies in Stellar Statistics. 1. Constitution of the Milky Way. First Memoir* (Lund, 1912) *LUÅ* n.f. afd. 2 vol 8 no 2 = *K. Fysiografiska Sällskapets Handlingar* n.f. vol 23 no 2); Malmquist, "Charlier".

Charlier used, in making the first study in stellar statistics, empirical data from published sources: the Carte du Ciel, the Bonner Durchmusterung, the Harvard Durchmusterung, the Harvard Revision and the Potsdam General-Catalog. He studied the brightness of stars in 48 zones of the sky. Pushing the data through the statistical apparatus Charlier concluded that the Milky Way had an extension of between 600 and 1400 siriometers.[191]

The next study dealt with the motions of stars: radial velocities and proper motions for stars in the same 48 zones as the first study.[192] A main result in this work is a criticism of Kapteyn's discovery of the stellar drift. Kapteyn's work on the motions of the stars had resulted in the idea of the stellar drifts: stars do not move randomly, but in two great drifts.[193] Schwarzschild had been critical of Kapteyn's result, arguing that two streams of stars ought to mean that other kinds of differences between two stellar populations should also be visible in material on spectra and brightness. Charlier joined in the criticism.[194]

Charlier next turned to stars of spectral type B.[195] He used a method for finding the distance of a number of stars of spectral type B by using data on the proper motion and radial velocity of the star.[196] He could then calibrate a distance scale to these stars. A star's distance can be found if the absolute magnitude (the brightness the stars would have if placed at a distance of one siriometer, in Charliers definition) and the apparent magnitude are both known. The absolute magnitude is, according to Charlier, closely tied to a star's diameter and effective temperature. Temperature is mirrored in the spectral class, with stars of the same spectral type having the same temperature. Only the largest stars reach the high tem-

---

[191] Charlier, *Constitution*, 55.

[192] C.V.L. Charlier, *Studies in Stellar Statistics II. The Motion of the Stars* (Lund, 1912-1913) (*LUÅ* n.f. afd. 2 vol 8 no 4 – *K. Fysiografiska Sällskapets Handlingar* n.f. vol 23 no 4).

[193] J.C. Kapteyn, "Statistical Methods in Stellar Astronomy", Howard J. Rogers ed., *Congress of Arts and Sciences, Universal Exposition, St Louis, Volume IV: Physics, Chemistry, Astronomy, Sciences of the Earth* (Boston and New York 1906), 396-425; Paul, *Statistical Cosmology*, 84-94.

[194] Paul, *Statistical Cosmology*, 126ff.

[195] C.V.L. Charlier, *Studies in Stellar Statistics III: The Distances and the Distribution of the Stars of the Spectral Type B* (Uppsala, 1916) (*Nova Acta Regiæ Societatis Scientiarum Upsaliensis* ser 4 vol 4 no 7 – *Meddelande från Lunds Astronomiska Observatorium* ser 2 no 14).

[196] C.V.L. Charlier, "Die Abstände der Sterne vom Spektral-Typus B", *AN*, vol 201 no 4801 (1915), 9-12.

peratures that are characteristic of the B stars; B stars can be thought of as having a rather small dispersion in absolute magnitude.[197] The distances of the B stars can hence be calculated from the known apparent magnitude.

To find out which stars are of the spectral type B, Charlier used the surveys of large numbers of stellar spectra that were made by Henrietta Leavitt at the Harvard observatory. He also used data on the star's apparent magnitudes and other data. The main result is a picture of the distribution of the B stars in space. From a well-defined centre, they thin out until the limits of the system is reached at about 200 siriometers from the centre; the sun is 18.2 siriometers from the centre. The system – three times as wide as it is high – is thought to be a 'skeleton' for the rest of the stellar system.[198]

In 1918, Charlier studied the distribution of clusters of stars.[199] Thinking that, statistically, clusters of stars were of about the same size, the apparent size of the objects was used as distance indicators, with small clusters viewed as distant clusters. He found that the clusters form a system of about the same shape as the one consisting of B type stars. He then assumed that the two systems coincided also as regards distances; he thus used the distance to the system of B stars as a yardstick for measuring the distance to the system of stellar clusters.

*

An infinite universe meant problems for the cosmologist. Olber's paradox is such a well known problem: given that the universe is infinite and stars are distributed with about the same density throughout the universe, the line of sight will everywhere reach the surface of a star. Wherever we look we would see a star; the sky would be shining like the surface of the sun. A similar paradox was formulated by Hugo von Seeliger, who argued that an infinite universe ought to contain infinite mass, thereby exerting an infinite force of gravity on us. One would have to give up the idea of the universe being infinite, or the homogeneity of Newton's law of gravity. A correction term could make gravitation somehow absorbed over long distances.[200]

---

[197] Charlier, *Spectral Type B*, 2.

[198] Charlier, *Spectral Type B*, 104.

[199] C.V.L. Charlier, *Studies in Stellar Statistics IV: Stellar Clusters and Related Celestial Phenomena* (Lund, 1918).

[200] Paul, *Statistical Cosmology*, 66-70.

Early in his career, Charlier had shown an interest in the questions of the large-scale structure of the universe. In a paper published in 1896, he argued that the spiral nebulæ where local objects inside our system, and Seeliger's paradox a strong indication that the universe could not be infinite. He argued, however, that the universe was eternal.[201] In 1908 Charlier published a paper where he argued for a way out of the dilemmas posed by Olber's and Seeliger's paradoxes, and for a universe that is infinite.[202]

Charlier argued that the nebulæ visible on the sky as small smudges of diffuse light consisted of immensely distant systems of stars. A minority of nebulæ were local objects inside the Milky Way system. Charlier argued that the distant nebulæ were the units of a next level, building up what he called a nebula of the second order, just like the stars build up our stellar system. This nebula of nebulæ is, together with other similar objects, part of a larger organisation and so on; the universe is fractal and infinite. The universe could be infinite (while still avoiding the paradoxes) if the sizes of the members making up the nebulæ fitted in what he called the convergence criterion of the cosmos.[203]

Charlier also tested the idea by calculating some aspects of the universe from the theory. The closest nebula of the second order turned out to be distant and invisible to the instruments of the day. The calculation for the level below that gave a result that was roughly compatible with the Andromeda Nebula's apparent size.[204]

In 1922 he returned to the question with a third paper. The convergence criterion was somewhat changed, but qualitatively the result was the same as in 1908.[205] He also discussed the distribution of nebulæ across the sky. Charlier and the staff at the Lund observatory had been working for some years on the statistics of some 13 000 nebulæ and clusters cata-

---

201 C.V.L. Charlier, "Ist die Welt endlich oder unendlich in Raum und Zeit?", *Archiv für systematische Philosophie* vol 2 (1896), 477-494.

202 C.V.L. Charlier, "Wie eine unendliche Welt aufgebaut sein kann", *Arkiv för matematik, astronomy och fysik* vol 4 no 24 (1908).

203 The term convergence criterion is from C.V.L. Charlier, "Huru en oändlig värld kan vara uppbyggd", *PAT* 1922, 1-8.

204 A nebula of the second order would have an apparent diameter of 0.2" and be of 37th magnitude. In Charlier's time, objects of about 19th magnitude could be observed.

205 C.V.L. Charlier, "How an Infinite World May be Built Up", *Arkiv* vol 16 (1922) no 22.

logued by Dreyer in the New General Catalogue.[206] The nebulæ were distributed in such a way in the sky that they seemed to avoid the Milky Way. This fact, known earlier and used as an argument against the nebulæ being galaxies since they somehow related to our system of stars, was explained by Charlier as an effect of obscuring material in the Milky Way system. This dust was concentrated to the plane of the Milky Way system and gave the effect that the nebulæ avoid the plane of the Milky Way.[207]

Statistical Practice in Science and Society

The statistical methods of Charlier and other stellar statisticians were of interest to scientists from other fields. The methods for handling large amounts of data of stars could also be used for the empirical material in other sciences, or in other parts of society. The statistical methods were an interface between astronomy and other sciences, between science and society.

The flow in towards Charlier's institute that was remarked upon above – a flow of practices and techniques, like calculating machinery, human computers, the mobility of observational data entailed by the photographic technology – corresponds with a flow in other directions. Not only were models of the Milky Way and the large-scale arrangement of the universe produced in his theoretical laboratory, but a method for analysis of other stuff than the radial velocities of stars or the properties of clusters of stars was also developed. Charlier viewed statistical methods as indispensable tools for handling reality in more rational ways. His statistical programme was a way to modernise society and the world we live in, a way to let the scientist provide rational answers to the questions of the day.

From the beginning of his studies in statistics, Charlier had a vision of statistics as something useful to the world outside of astronomy. One aspect of this was his work to found the field in the university. Since 1902

---

[206] C.V.L. Charlier, "Preliminary Statistics of Nebulæ and Clusters", *Arkiv* vol 9 (1913) no 15.

[207] C.V.L. Charlier, "How an Infinite World May be Built Up", 33; "Huru en oändlig värld kan vara uppbyggd". This is one point where Charlier's cosmology differs from the work of Kapteyn, who had argued against interstellar absorption. E.R. Paul, "Kapteyn and Statistical Astronomy" in Hugo van Woerden, Ronald J. Allen & W. Butler Burton eds., *The Milky Way Galaxy* (Dordrecht, 1985), 25-42, 33.

there existed a chair of statistics and political science at Lund, a position held by Pontus Fahlbeck. Strongly critical of the conception of statistics held by Fahlbeck and others, Charlier viewed it as an elderly form of statistics, grounded in earlier work on censuses and descriptions of states. The very word was derived from the Latin *status*, which during the middle ages began to mean country.[208] That type of statistics was criticised by Charlier, who argued that more mathematical analysis was needed in statistics. Charlier wanted to lessen the tabulating of societal facts and instead bring more mathematics into lundensian statistics. He argued that the future of statistics hinged on the subject being able to obtain more mathematical rigor.

In arguing for a renewal of statistics he used utilitarian arguments. Modern statistical practices were obviously needed for the state in its process of modernisation. Charlier identified a change, a new kind of society had come into existence where rational arguments and control enforced by statistical methods and data were necessary. This placed his own skills as well as those of his pupils in a position to take part in an important area of society. Sometime after Sven Wicksell had got his PhD in 1915, Charlier argued for a research grant for Wicksell. This was important, since the situation in Lund was in an "emergency": Fahlbeck had retired and Lund did not have any representative "for scientific statistics". He also urged university officials to look at the issue's

> important practical benefits to the state. Earlier, hardly any department of state [*ämbetsverk*] existed where statistical methods were used; now one large department after another has arisen with hundreds of positions, for which knowledge of mathematical statistics is or should be an indispensable part of the skill of the civil servants working there. I might point at the Royal Board for Social Issues [*K. Socialstyrelsen*], the National Board of Insurance [*Riksförsäkringsanstalten*], the Pension board [*Pensionsstyrelsen*], the recently decided upon Office for Accidental Insurance [*Olycksfallsförsäkringsanstalten*], not to mention all the similar boards in regional government and private companies. If ever the needs of education [for practical matters] shall be taken into account, this is the time.[209]

[208] Wynnard Hooper, "Statistics", *Encyclopædia Britannica* 11th edition (New York, 1911) vol 25, 806-811, 806.

[209] Charlier's report arguing for a research grant [*docentstipendium*] for Sven Wicksell, n.d. (c. 1916 because of the chronology of the papers), Astronomy department "Koncept, institutionshandlingar 1907-1928", 635f, LUB.

For Charlier, it was important to differentiate between mathematical statistics and other forms of statistical reasoning. When it was discussed whether to institute a new chair of statistics at Lund University, a committee was formed, where Charlier argued for mathematical statistics. The result was that the committee issued a statement that the previous combination of statistics and political science used at the Lund university (Fahlbeck's chair) was too wide.[210] Eventually, the chair in statistics at Lund was held by Charlier's pupil Sven Wicksell. Charlier and mathematical statistics made inroads in the multidisciplinary field of statistics.

*

Statistics served as an interdisciplinary field, a trading zone, where Charlier's expertise was, for a time, wanted by practitioners from other disciplines. Interest came, not the least, from medicine and biology; as is evident in the Charlier correspondence. His writings did really reach a wide audience and the subject was evidently perceived as something that might be of use in various fields. Georg Henrik Dovertie, a leading doctor of medicine in Gothenburg and Bohus counties (*förste provinsialläkare*) was interested in applying the statistical methods in medicine. Hans Valdemar Gertz, active at the Karolinska institute of medicine and a doctor that specialised on the physiology of vision, wrote Charlier after having read his works in statistics. Gustaf Emanuel Hultkvist, second city doctor of Stockholm, wrote Charlier to ask for assistance in how to use statistics for studies of the correlation of mortality and various diseases. The agricultural chemist Carl von Scheele had published a study of the amount of starch in potatoes, using Charlier's handbook *Grunddragen av den matematiska statistiken*. Scheele thought that the book might come to use as a textbook in teaching statistics, when farming schools began teaching the subject.[211]

Interest came also from abroad. Pure statisticians, as well as workers in other fields, who were interested in applying statistics to their line of work, wrote Charlier. His works were requested at the National Institute for Medical Research at the British Medical Research Council. Werner

[210] "Till humanistiska sektionen", Astronomy department "Koncept, institutionshandlingar 1907-1928", 665f, LUB.

[211] Georg Henrik Dovertie to Charlier 6 September 1915; Hans Valdemar Gertz to Charlier n.d.; Gustaf Emanuel Hultkvist to Charlier 28 November 1915; Carl Arvid Schultz-Steinheil to Charlier, all in Charlier's correspondence, LUB.

Cajanus, who was working in forestry in Helsingfors was interested in studying how very old populations of spruce were dispersed as regards size during different stages in their life. He had made use of Charlier's paper "Researches into the theory of probability". A similar interest had Wolff von Wülfing, active at a forest research institute in Java.[212]

An interest in statistics linked Charlier to the United States. Lundensian (or Scandinavian) statistics were thought of as one of a few centres of statistical science. Arne Fisher, a statistician with a Danish background, was working as statistician at the American Telephone and Telegraph Company and The Prudential Insurance Company of America. When Malcolm Rorty, head of statistics at AT&T, said in 1920 that the company needed more statisticians, Fisher suggested that Rorty should attempt to get statisticians educated at Lund, since the university had the best education for statisticians.[213]

In 1916 a school of public health opened at Johns Hopkins University. A department of biometry and vital statistics was connected to it, the head of which was the population biologist and statistician Raymond Pearl.[214] Pearl, who had studied under the statistician Karl Pearson, had showed an interest in Charlier's paper on the *Trientalis Europæa* before the war.[215] He held the work of the Lund school of statistics in such high esteem that he wanted to send his graduate student H.E. Niles there to study under Charlier or Wicksell; Charlier replied that he could let Niles come to Lund in 1922.[216] Several others were interested in going from the US to study under Charlier.[217]

---

[212] Anon. at the Medical Research Council to Charlier, 3 November 1920; Werner Cajanus to Charlier 14 November 1912, 17 June 1913, 30 June 1913, 14 October 1913, 15 November 1913; H.E. Wolff von Wülfing to Charlier 3 March 1923 – all in Charlier's correspondence, LUB.

[213] Arne Fisher to Charlier, n.d.; Malcolm Rorty to Charlier 21 September 1920 and 23 December 1920 – all in Charlier's correspondence, LUB. Charlier did not find a statisticians willing to work in USA for $2400 per year. Edwin B. Wilson to Charlier 17 March 1923, Charlier's correspondence, LUB.

[214] Pearl to Charlier 27 January 1914, Charlier's correspondence, LUB; "History of the Johns Hopkins Department of Biostatistics",
http://www.sph.jhu.edu/biostats/history.html 17 October 1996.

[215] Pearl to Charlier, 27 January, 25 February, 1914, Charlier's correspondence, LUB.

[216] Pearl to Charlier 22 February, 15 May 1921, Charlier correspondence, LUB.

[217] Burton Howard Camp to Charlier 4 December 1922, 7 February 1923; James Waterman Glover to Charlier 29 March 1924; Paul R Rider to Charlier 21 September 1927 – all in Charlier's correspondence, LUB.

In the 1920's a department of vital statistics at the school of public health was founded at Harvard. Edwin B. Wilson was active there, interested in applying the use of statistics in medicine. In writing to Charlier, he claimed the Swedish school's statistics to be more applicable to biology and medicine than Karl Pearson's methods. Wilson hoped that Charlier would someday come to the US and lecture.[218]

Charlier thought his statistical works were of such good quality that he did not hesitate to try to reach an international audience. Before the war, he had tried to get his statistical writings published in English but the publishers he approached were negative for reasons of economy.[219] Charlier tried to get *Grunddragen av den matematiska statistiken* published in the *Edinburgh Mathematical Tracts* in 1918, but failed.[220]

After the war attempts were made to make Charlier's works available in other languages. In 1920, the introductory textbook on statistics was published as *Vorlesungen übder die Grundzüge der mathematischen Statistik*; by 1931, it had reached a third edition. Edwin W. Kopf, statistician at the Metropolitan Life Insurance Company was interested in translating Charlier's writings to English "in order that it may have a wider circulation in our colleges and universities and among the statistical offices." Floyd F. Burtchett, a fellow of economics and statistics at the University of Wisconsin, had made a translation of the *Grundzüge der mathematischen Statistik* to English, and wanted Charlier's permission to have it published. The Macmillan company – approached by Charlier, Kopf and Burtchett, turned town the proposition: books for teaching at the undergraduate level in the US should contain numerous examples of applications and practical training.[221]

---

[218] Edwin B. Wilson to Charlier 25 October 1922, 17 March 1923, Charlier's correspondence, LUB.

[219] The Macmillan & Co to Charlier 15 February 1910, 18 July 1912; Cambridge University Press to Charlier 7 March 1911; Oxford University Press to Charlier 20 July 1912 – all in Charlier's correspondence, LUB.

[220] Edmund Taylor Whitaker to Charlier 3 April 1918, Charlier's correspondence, LUB. Whitaker explained the negative responce to economical problems because of the war; he would come back to Charlier when funding was better, which he (it seems) never did.

[221] Edwin W. Kopf to Charlier 30 January 1923; Floyd F. Burtchett to Charlier 13 February 1923; The Macmillan company to Charlier 7 April 1924 – all in Charlier's correspondence, LUB. Later still, Charlier did get an introduction to statistics in astronomy published in French. C.V.L. Charlier, *Application de la théorie des probabilités a l'astronomie* (part of *Traité du calcul des probabilités et de ses applications*) (Paris,

*

Charlier did not hesitate to promote the use of statistics in areas outside of academia. His vision of a utilitarian use of statistics was enhanced by his urge for a rational and truly scientific way of running society. To this end, he participated as an expert in several state committees, and saw some of his students make quite distinguished careers in fields rather remote from the rarefied airs of theoretical astronomy. (The point being made here is, of course, that that air never is that rarefied.) Charlier and his students thus played a part in the more rational way of running the state that made an impact during the late 1800's and early 1900's.

In this respect, Charlier and his pupils fit in with a trend in Swedish science. Historians of science Gunnar Eriksson and Thomas Kaiserfeld have studied scientists and their participation in Swedish society during the decades surrounding 1900. They worked in industry, in state departments, and in insurance companies, played the role of experts, and moved in a network of communication between state, industry, and the universities and research institutes.[222]

In 1898 Charlier participated in an state committee that analysed the pricing of train tickets in Sweden. The schemes for the cost of train tickets were about to be reformed, and a rational study of the effects of the new schemes proposed was needed.[223] Further on, his abilities to handle the changes of life statistically were put to use when he analysed several pension schemes.[224] One task concerned the pension schemes for relatives of deceased employees at the Lund university. First, Charlier set out to discuss data on the lives of teachers and scientists employed at Lund University; he used a number of membership directories that chronicled the

---

1931), and *Grunddragen af den matematiska statistiken* was translated as *Vorlesungen über die Grundzüge der mathematischen Statistik* (Lund, 1920)

[222] Thomas Kaiserfeld, *Vetenskap och karriär*, chapter 5; Gunnar Eriksson, *Kartläggarna*, chapters 4-7.

[223] *Betänkande afgifvet den 1 oktober 1898 af komitén för uppgörande af förslag till ny taxa för personbefordringen å statens jernvägar* (Stockholm, 1898).

[224] For the pension society "Assuransföreningen 401" in Landskrona he calculated the probable deaths of its members 50 years in to the future. C.V.L. Charlier, *Sannolikhetsberäkningar för assuransföreningen 401 Landskrona* (Landskrona, 1907). He made calculations for a pension society of military officers. John Källström of "K. Skånska trängkårens underofficerskår" to Charlier, 19 November 1914, Charlier's correspondence, LUB.

lifetimes of the university teachers. He tabulated data on birth, time of employment at the university, death, year of marriage, the wife's (or wives') year of death and birth, children and so on.[225] Charlier could give guidelines for the amounts that could be paid without risking the collapse of the pension system.

After the war, a state lottery was discussed. The government gave Charlier the assignment to "investigate the existence of state lotteries in foreign countries, their organisation and the experiences made in connection with them." Charlier took the task seriously. He took a sabbatical leave from the observatory, made study trips to several countries, and corresponded with personnel at Swedish embassies and so on.[226]

Charlier was positive towards starting a lottery. It would be possible to construct a lottery so that it would be a source of income for the state, and at the same time educate Swedes about the laws of statistical variation. It would, furthermore, be possible to construct a lottery in such a way that its most negative effects would be minimised, such as poor people spending all their money on gambling. His study led to no immediate results, but when a state lottery eventually was constructed, it was on the lines suggested by Charlier.[227]

Some of the people that studied for Charlier became astronomers, some became teachers, and some became statisticians. The Lund way of doing astronomy, the statistical school around Charlier, provided astronomers with a career alternative. They could become statisticians proper, or work as consulting experts on statistical matters. In this way, they participated in the growth of the Swedish public sector, which was in need of people that could help planning society. The astronomers from Lund could co-operage in this venture. Under Charlier they had learnt a set of practices that could be used in domains outside of astron-

---

[225] C.V.L. Charlier, "Utredning angående pensionering av änkor och barn efter ämbets- och tjänstemän vid Lunds universitet", typewritten MS dated 21 September 1923, Lund observatory library. The data could at first seem odd. Mean longevity of university teachers had not changed from the 1660's to the 1920's, despite it having risen in society at large. Charlier attributed the change in society at large to better hygienical standards, whereas university professors had been on the same level all the time. Charlier, "Pensionering", 5ff.

[226] "Sammandrag af K.M:ts brev angående lotteri-sakkunnig", Astronomy department "Koncept, institutionshandlingar 1907-1928", 292, LUB; Lund observatory yearly report *LUÅ* 1921-1922, 31f; C.V.L. Charlier, *Utredning angående inrättandet av ett svenskt statslotteri* (Malmö, 1922) (SOU 1922:24).

[227] C.V.L. Charlier, *Svenskt statslotteri*, 96ff; Malmquist, "Charlier", 400ff.

omy. They had learnt to handle statistically the numbers of sunspots, the motions and distances of stars; these methods were also suited to handle phenomena like fertility and mortality, or in the field of insurance. Insurance is a key concept: statistics could be used in the drive towards more and more possibilities of insurance against unemployment, damages on buildings and so on built up by the public sector. Stellar statistics could be – and was – generalised as a mobile competence that the astronomers could bring outside of the astronomical community. Several of Charlier's pupils ended up in places far away from academic astronomy; with them, the awareness and use of the statistical methods also spread. Therefore, Charlier and Lund observatory under his reign, has a place in the history of statistics, as well as a place in the history of the modern Swedish state.

The number of pupils that continued in astronomy was slightly less than a third. This is a situation that probably has been rather common throughout the period studied here; not all astronomers have managed to stay in academic astronomy and have instead moved on to being teachers. Many PhD students probably had no scientific career in sight when they began their studies, but rather aimed for a *lektorat* at a secondary school, a post for which a PhD was necessary. One way to stay in astronomy has been to move abroad. The emphasis on statistical methods by Charlier opened up new possibilities for astronomers that could not get positions inside academic astronomy. After a time, two professors of statistics in Sweden were pupils of Charlier's: Josua Linders in Uppsala and Sven Wicksell in Lund. Several other pupils took part as statistical experts in the rapidly expanding field of state-run commissions planning and surveying the society. Statistical experts were needed; the number of commissions increased. The time of Charlier's interest in statistics in astronomy coincided with a rapid rationalisation of Swedish society, an urge to plan life, to manage society in a rational way.[228] Between 1855 and 1904 531 state committees were installed, between 1905 and 1954 the number was 2 729.[229] In these, statistical data and analysis was often used.

---

[228] Thorsten Nybom, "Samhällsformation och samhällsorganisation i Sverige 1890-1975 – en principskiss", in Thorsten Nybom & Rolf Torstendahl eds., *Byråkratisering och maktfördelning* (Lund, 1989). There were predecessors to Charlier. During the eighteenth century, a leading statistician was the astronomer Pehr Wilhelm Wargentin. Karin Johannisson, *Det mätbara samhället: Statistik och samhällsdröm i 1700-talets Europa* (Stockholm, 1988), 159ff.

[229] Hans Meijer, *Kommittépolitik och kommittéarbete: Det statliga kommittéväsendets utvecklingslinjer 1905-1954 samt nuvarande funktion och arbetsformer* (Lund,

One of Charlier's pupils that advanced far in state planning and control work was Olof Åkesson.[230] He had written a PhD thesis on the statistics of sunspots, based on a photographic material from the Greenwich observatory. In 1916 he started working for the State Insurance Board [*Riksförsäkringsanstalten*]. He worked in several important committees: the 1921 committee on unemployment insurance, 1926 committee on unemployment and the 1928 study on retirement insurance.[231] In 1930 he became director of the office of insurance inspection, a state board that was organised by the ministry of trade. Its task was to oversee the activities of insurance companies. During the 1930's and 1940's he worked on a number of committees that analysed unemployment, retirement schemes and insurance issues.

Carl Filip Lundahl studied the properties of stars of spectral type F in his PhD thesis. Then he worked a short time as assistant at the State Insurance Board. Charlier tried to get him involved with the eugenics institute at Uppsala, run by Herman Lundborg, but that did not happen (Lundahl instead became a teacher in Jönköping).[232] Instead, Josua Linders, another of Charlier's pupils, got a place at the eugenics institute at Uppsala.

Linders had been assistant (*amanuens*) at the Lund observatory for about two years and finished a licentiate degree, when he started at the Central bureau of statistics in 1912, were he worked until 1917.[233] Positions as statistician in a sugar manufacturing company and the central board of school planning followed, before Linders got a position at the eugenics institute at Uppsala in 1921.

1956), 8. Navigation through the large literature issued by the committees is made easier by use of *Förteckning över statliga utredningar 1904-1945 utgiven av riksdagsbiblioteket* (Norrköping, 1953).

[230] Biographical data on Åkesson in *Svensk uppslagsbok* second edition (1947-1955).

[231] *Betänkande och förslag ang. offentlig arbetsförmedling och statsbidrag till arbetslöshetskassor* (Stockholm, 1923) (SOU 1922:59); *1926 års arbetslöshetssakkunniga. Betänkande och förslag ang. arbetslöshetsförsäkring, arbetsförmedling och reservarbeten* (Stockholm, 1928) (SOU 1928:9); *1928 års pensionsförsäkringskommitté. Betänkande med förslag rör. revision av den allmänna pensionsförsäkringen* (Stockholm, 1934) (SOU 1934:18).

[232] Lundahl to Charlier, 22 February 1922, Charlier's collection, LUB.

[233] Sten Malmquist, "Linders, Frans Josua", *SBL* and the article on Linders in Thoralf Fries & Ernst von Döbeln, *Uppsala universitets matrikel höstterminen 1936* (Uppsala, 1937).

Linders stayed at the eugenics institute until 1927, when he left after a quarrel with the director, Herman Lundborg.[234] Linders' statistical expertise was a vital part of the institute's survey of the population in racial terms. The population was divided into what was called the purer Nordic type, the purer east Baltic type, and the predominantly dark types. On the nordic conference for eugenics and anthropology in Uppsala in 1925, Linders demonstrated the antropometric studies, as well as the use of a Hollerith machine for automatic sorting and tabulation of large amounts of numerical data.[235] Statistical expertise and number crunching abilities were important for the anthropological survey of Sweden. Linders not only dealt with statistics as a supporting science but also took part in the eugenic mapping of Sweden. Linders and Lundborg co-authored *The Racial Characters of the Swedish Nation*, and Linders himself wrote the study *Zur kenntnis der Kopfmasse in Schweden*.[236]

During his years at the eugenics institute, Linders completed a PhD thesis in statistics, on demographic studies of the Swedish clergy. A grant enabled him to do research at Uppsala University. He published on several statistical subjects: statistics of religion, the sizes of infants, shrinking nativity. In 1931 he was appointed professor of statistics at Uppsala University, after having worked in a leading position in the census of 1930. He also supported several committees with statistical knowledge in matters of education, defence and taxes.

Sven Wicksell - son of the economist Knut Wicksell - finished a PhD degree under Charlier in 1915.[237] It dealt with the velocities of stars, but he soon worked on more exclusively statistical subjects. Soon after he had completed his PhD thesis, Wicksell became docent in mathematical sta-

---

[234] For the history of the eugenics institute, see Gunnar Broberg, *Statlig rasforskning: En historik över rasbiologiska institutet (Ugglan: Lund Studies in the History of Science and Ideas*, 4) (Lund, 1995). Eugenics in Scandinavia is covered in Gunnar Broberg and Nils Roll-Hansen, eds., *Eugenics and the Welfare State : Sterilization Policy in Denmark, Sweden, Norway, and Finland* (East Lansing, 1996).

[235] Broberg, *Rasforskning*, 38.

[236] Hermann Lundborg & Frans Josua Linders, *The Racial Characters of the Swedish Nation* (Uppsala, 1926); Frans Josua Linders, *Zur Kenntnis der Kopfmasse in Schweden* (*Lund Medd.* Ser II no 50A) (Lund, 1927) (also in *Festskrift tillägnad C.V.L. Charlier på hans sextiofemårsdag den 1 april 1927* (Lund, 1927).

[237] Knut Lundmark, "Wicksell, Sven Dag", *SMoK*; "Wicksell, Sven Dag", *Lunds universitets matrikel läsåret 1924-1925* (Lund, 1925); Thorild Dahlgren, "Sven Dag Wicksell", *Vetenskaps-societetens i Lund årsbok* 1939 (Lund, 1939), 151-161; Walter Gyllenberg, "Sven Dag Wicksell", *Kungl. Fysiografiska sällskapets i Lund förhandlingar* vol 9 (1939), 40-45.

tistics, the first such in Sweden. In 1920 he got a post as teacher in the subject. From 1918 he headed a statistical seminar at the Lund university, where 20 to 30 students participated each semester, to learn the theories and practices of statistics.[238] He also taught statistics at the university college in Gothenburg. In 1926 he was called to the new chair of statistics in Lund.

Wicksell's focus was on population statistics. He studied the nativity, and was worried about the diminishing numbers of births. He took part in the large committee on population, where his expertise was used in a report on sexuality and nativity in 1935.[239]

*

Stellar statistics was centered on the analysis of quite large amounts of data on stellar brightness, position, radial velocities etc. It was a kind of astronomy that tried to get to the outline of the Milky Way system by collecting data for as many stars as possible. This mass analysis was parallel with a mass-production of data, begun with the introduction and widespread adoption of photographic methods. Stellar statistics also took advantage of the mobility inherent in the stellar photographic material. Photographic plates could be removed from the site where they had been exposed and analysed elsewhere. This separation of the observation and analysis led to the construction, at Lund, of an astronomical activity that dealt almost exclusively with observations collected at other observatories. The Lund group used statistical methods to get at the structure of the stellar system based on published data or photographic plates.

The theoretical astronomers needed to make analysis of data more industrial; the point of the statistical analysis was that it needed large datasets that were to be analysed statistically. A female office workforce and the expansion and mechanisation of the office workspace presented Charlier with one possible solution to his number-crunching problems; the computers at the observatory were more or less ready to start work

---

[238] Yearly reports of the statistical seminar in *LUÅ* 1918 and onwards. Texts by Charlier, Yule, Arne Fisher, Wicksell and others were used. "Studieplan i matematisk statistik vid Lunds universitet 1917-1918), *Svenska aktuarieföreningens tidskrift* vol 4 (1917), 187f.

[239] Sven Wicksell, "Individens fruktsamhet och släktets förökning", *Ekonomisk tidskrift* 1930; *Betänkande i sexualfrågan avg. av befolkningskommissionen* (Stockholm, 1936) (SOU: 1936:59).

without extensive astronomical training, they were working like secretaries. Charlier and his stellar statistical programme were part of a larger process of transformation of society.

The statistical methods developed in Lund were presented to other sciences and were also forming the core of a statistical science in itself. Charlier's pupils Wicksell and Linders went to chairs in statistics at Lund and Uppsala. Statisticians trained in Charlier's statistical laboratory thus went on to present their capabilities to other areas of public life that needed statistical treatments: pension schemes, insurance policy, population studies, defence, education policy represents some of the public sector fields where statisticians from Charlier's lab were working.

LUNDMARK AND THE LUND OBSERVATORY

KNUT Lundmark (1889-1958) began his publishing career in classical astronomy with a paper published in 1917 on the orbit of the comet of 1802.[240] After that, however, his work belonged to more modern aspects of astronomy. He studied astronomy at the Uppsala observatory, with Östen Bergstrand as teacher. Bergstrand's interests in astronomy were at this time more focused on the modern applications of photographic astronomy, and he had started to make use of photography for determining colours of stars. The method of effective wavelengths had partly been developed by Bergstrand in collaboration with Ejnar Hertzsprung (see chapter two above). Bergstrand's pupils Knut Lundmark and Bertil Lindblad were beginning to work in that field opened up by their professor.

Lundmark and Lindblad began measuring the effective wavelengths of spiral nebulæ and globular clusters. Lindblad moved more and more in the direction of studying stars, and published several papers on the effective wavelengths of stars together with Bergstrand. Lundmark was more

[240] The major work done on Knut Lundmark is Anita Sundman, *Den befriade himlen: Ett porträtt av Knut Lundmark* (Stockholm, 1988); Anita Sundman, "Lundmark, Knut Emil", *SBL*; Robert Smith, *The Expanding Universe*, 105ff. Lundmark's place among religious scientists during the inter-war period in Sweden is discussed by Kjell Jonsson, "Physics as Culture: Science and *Weltanschauung* in Inter-War Sweden", in Svante Lindqvist ed., *Center on the Periphery: Historical Aspects of 20th-Century Swedish Physics* (Canton, MA, 1993). See also Martin Johnson ed., *Knut Lundmark och världsrymdens erövring – en minnesskrift* (Göteborg, 1961). Knut Lundmark, "Bestimmung der Bahn des Kometen 1802", *Arkiv för matematik, fysik och astronomy* vol 12 (1917), no 17.

interested than Lindblad in applying the method to nebulæ.[241] The method of measuring effective wavelengths made it possible to study the astrophysically interesting aspects of spiral nebulæ by using such small instruments as the ones that were available to Swedish astronomy.

Lundmark chose to continue with the nebulæ and clusters for his PhD work. A first study appeared in a paper in *Astronomische Nachrichten*, published in 1919.[242] During the spring of that year he finished the thesis and sent the manuscript to the Royal Academy of Sciences in May. Because of a strike among typographic workers in the summer of that year, it did not appear in print until the beginning of 1920.[243]

The work focused on the various observational results that existed concerning the place of the spiral nebulæ in the cosmos. Even if he used some observations made by himself with the 15 cm Uppsala astrograph, most of the material discussed was taken from the published literature.

Lundmark discussed the distances to the globular clusters of stars. He related the work of Shapley, who had determined distances to 69 globular clusters, and the work of Charlier. Shapley argued that the globular clusters provided a "skeleton" for our stellar system with much larger dimensions than the models proposed by the stellar statisticians. Shapley argued, in a series of papers published 1917-1918, that the stellar system had a diameter on the order of 300 000 light years, with the sun some 45 000 light years from the centre. This was a radical departure from the stellar statisticians' models, with dimensions around 10 000 light years and the sun placed near the centre. Shapley discarded, at the time, the island universe theory.[244] Charlier's distances to the clusters, and hence his size for the stellar system, were smaller by a factor of 40 to 290 times. Lundmark sided with Shapley in arguing for a large stellar system.[245]

The major part of the dissertation is a discussion on the distances of spiral nebulæ, using several distance indicators, of which the novæ were of particular interest. One prominent nova had been observed in the An-

---

[241] Hearnshaw, *Measurement*, 176ff; Lindblad to Lundmark 3 July 1919, Lundmark's collection, LUB.

[242] Knut Lundmark, "Die Stellung der kugelförmigen Sternhaufen und Spiralnebel zu unseren Sternsystem: Ein Versuch, ihre Parallaxen zu schätzen", *AN* vol 209 (1919), 369-380.

[243] Knut Lundmark, "The Relations of the Globular Clusters and Spiral Nebulæ to the Stellar System: An Attempt to Estimate their Parallaxes", *KVAH* vol 60 no 8 (1920), xv.

[244] Robert Paul, *Statistical Cosmology*, 191-202.

[245] Knut Lundmark, "Globular Clusters and Spiral Nebulæ", 4, 20.

dromeda nebula in 1885. In 1917 Ritchey at Mount Wilson discovered a nova in the spiral nebula NGC 6946. Curtis at Lick soon found other novæ. Lundmark thus had access to a total number of 18 novæ for the discussion in the thesis.[246] The distances to a number of local novæ had been determined by various methods. Under the assumption that the novæ in the Andromeda nebula were similar to the local objects, the distance to the nebula could be determined from the knowledge of local novæ's absolute magnitude. Lundmark tried to calibrate a distance scale with novæ as the yardstick. The main result was that spiral nebulæ *could be* systems at immense distances. Lundmark estimated the distance to the M31, the Andromeda Nebula, to around 600 000 light years.[247] This is how Lundmark's result could be interpreted, Lundmark himself had a quite diplomatic tone in the thesis, perhaps influenced by the cautious Bergstrand. He did not want say for sure that the nebulæ were distant stellar systems. Lundmark summarized his findings thus:

> The present investigation has given as the main result that the spiral nebulæ must be considered as situated at considerable distances from the solar system. Whether they are Jeans' star-producing mechanisms or remote galaxies is, on the other hand, more difficult to decide. Possibly we might in the present facts see a suggestion that the latter is the case, but the spiral nebulæ do not, however, seem to be of such dimensions as those that should be ascribed to the galactic system with regard to Shapley's investigations, and much also speaks against regarding the Galaxy as having a structure, analogous to that of the spiral nebulæ.[248]

In an appendix to the thesis, dated 26 January 1920, Lundmark introduced a cautious remark as to the large galaxy proposed by Shapley. "Several other facts are also indicating that the distance to the galactic clouds cannot be regarded to be as great as is demanded by assuming the globular clusters as being related to the Milky Way [...] it appears questionable that whether already 100 000 light-years is not too great a value for its diameter."[249]

Defence of the dissertation took place on the 21 of February 1920 with Bergstrand as opponent. Reactions from the colleagues in Sweden were mixed. The Charlier school of stellar statistics was, at least numer-

[246] Michael A. Hoskin, "Ritchey, Curtis and the Discovery of Novæ in Spiral Nebulæ", *JHA* vol 7 (1976), 47-53; Lundmark, "Relations", 53.

[247] Knut Lundmark, "Globular Clusters and Spiral Nebulæ", 59.

[248] Knut Lundmark, "Globular Clusters and Spiral Nebulæ", 62.

[249] Knut Lundmark, "Globular Clusters and Spiral Nebulæ", 78.

ary, the leading group of stellar astronomers and their models were about ten times smaller than the ones proposed by Lundmark. Charlier, whose results were at stake, wrote a polite letter, thanking him for the copy of the thesis and wishing him good luck.[250] Bergstrand thought it was somewhat regrettable that Lundmark disagreed with Charlier, "but that can not be helped. You have of course been very cautious in your statements and precisely considered every word. I don't think that Charlier will become unpleasant towards you in any way because of this".[251] Bergstrand was positive, giving the thesis a good grade in opposition to Hugo von Zeipel, and he wrote positively of it in *Nordisk Astronomisk Tidsskrift*.[252] Bertil Lindblad, who had been Lundmark's fellow graduate student at Uppsala, was unsure of where the nebulæ were situated in the cosmos. He saw the rapid rotations of some nebulæ that Adrian van Maanen had measured as a considerable stumbling block for Lundmark and other proponents of the island universe theory.[253]

Lundmark used publications in popular science to spread his results, and here he was not as cautious. In the spring of 1921 he wrote the article "Världsalltet" (the cosmos) for the second edition of the encyclopædia *Nordisk familjebok*. Lundmark argued here that previous models of the cosmos had been wrong: the spiral nebulæ were distant systems of stars, instead of local objects. Bergstrand and Lundmark were originally supposed to have written the article together, but Bergstrand withdrew, as he could not support the contents. Lundmark used too much of his own results in the article and with far too much detail. Bergstrand and Lundmark exchanged several letters on this issue.[254]

Something similar occurred when Lundmark wrote the small book *Spiralnebulosorna: Ett nutida astronomiskt forskningsproblem*.[255] The criti-

[250] C.V.L. Charlier to Knut Lundmark 27 February 1920, Lundmark's correspondence, LUB.

[251] Östen Bergstrand to Knut Lundmark 15 July 1919, Lundmark's correspondence, LUB.

[252] Östen Bergstrand to Knut Lundmark 12 November 1921, Lundmark's correspondence, LUB; Östen Bergstrand, "Världsalltets utsträckning", *Nordisk Astronomisk Tidsskrift* ny följd vol 1 (1920), 33-40.

[253] Bertil Lindblad to Knut Lundmark 27 October 1920, Lundmark's correspondence, LUB.

[254] Östen Bergstrand to Knut Lundmark 12 June, 30 August, 6 September, 12 November 1921, Lundmark's correspondence, LUB.

[255] Knut Lundmark, *Spiralnebulosorna: Ett nutida astronomiskt forskningsproblem* (Studentföreningen Verdandis småskrifter no 247) (Stockholm, 1922).

cism was the same this time: subjectivity and too much detail. Bergstrand thought that Lundmark gave too much room for preliminary results that had not yet stabilised, as well as his own results.[256] Lundmark as writer of popular science differs a lot from Bergstrand; the latter thought that such literature ought to mirror a scientific consensus. In popular science, there was no room for the author's own results or fresh, untested results. Security and clarity was what Bergstrand aimed at. Lundmark was more interested in making use of popular writings to further his own ideas in matters that had not reached closure.

Lundmark and American Astronomy

Communication patterns in Swedish astronomy were changing: more and more it became oriented westward (see chapter six). In June 1919 the Sweden-America Foundation had been founded on initiative from a number of people in academia, industry, and culture that wanted to strengthen the ties between Sweden and USA.[257] What was perhaps considered the finest of the foundation's grants had been donated by the artist Anders Zorn for Swedes that wished to do scientific studies in the US. In 1920 the grant was awarded to Knut Lundmark.[258]

Lundmark spent two years in USA, where he got the chance to work at several leading observatories, and to make contacts with American colleagues. He spent most of the time at the two large observatories in California, Mount Wilson observatory and the Lick observatory on Mount Hamilton.

The astronomers in California were aware of Lundmark's thesis before he arrived. Harlow Shapley and Adriaan van Maanen had read it thoroughly.[259] Shapley had written a letter to Lundmark before he went to the US, in which he welcomed him to the field of cosmology, and

---

[256] Östen Bergstrand to Knut Lundmark 30 August 1921, Lundmark's correspondence, LUB.

[257] Among the names are such scientists and cultural stars as Svante Arrhenius, The Svedberg, Selma Lagerlöf, Anders Zorn, Torgny Segerstedt, Hjalmar Lundbohm, Jacob Wallenberg, Ivan Bratt, Hjalmar Branting, Dan Broström and Nathan Söderblom. Dag Blanck, *Sverige-Amerika stiftelsen: De första sjuttio åren 1919-1989* (Stockholm, 1989), 6f.

[258] Dag Blanck, *Sverige-Amerika*, 14f.

[259] Bertil Lindblad to Knut Lundmark 27 October 1920, Lundmark's correspondence, LUB.

made remarks that were quite critical of some of Lundmark's results, for instance the immense distances of the nebulæ. He praised a passage in the thesis as being "sound and conservative" and stated that he hoped "that we won't be led into prejudiced and weak interpretations by our hypotheses". Shapley disagreed with Lundmark's results and was critical of his style, hinting that the results were loosely constructed.[260] In a paper published in late 1921, Lundmark discussed the great spiral nebula in Triangulum, M33. The paper was rather short and had a clear message of where Lundmark stood on the issue of the spiral nebulæ; the diplomatic and careful tone of the PhD thesis was not present.[261] In the paper Lundmark could augment arguments from his thesis with observations secured at the Lick observatory. He noted that a spectrogram he had taken with Lick's Crossley telescope – with up to 30 hours of exposure – indicated the contents of M33: "From the spectral evidence it seems probable that the spiral nebula consists of ordinary stars, clusters of stars and some nebular matter."[262] In his results, Lundmark clashed with the views of several astronomers. He wrote home to Bergstrand that his spectrogram gave a good indication that there were stars in the nebula M33 rather than gas, but he also noted to his cautious teacher Bergstrand that it was still uncertain how large the spirals were, if they were on a par with the Milky Way system.[263]

Harlow Shapley wrote to Lundmark in late December to inquire about one of the major stumbling blocks for the island universe hypothesis: "May I ask why you ignore van Maanen's measures of rotation in Messier 33 and other spiral nebulæ when you discuss the problem of their distances and constitution?"[264] The measures by van Maanen gave to many astronomers strong indications that the nebulæ could not be stellar systems at very large distances, but rather were rotating clouds, perhaps solar systems in the making. In a later letter, Shapley identified Lundmark as the strongest supporter of the island universe theory, and again

---

[260] Harlow Shapley to Knut Lundmark May 1920 (no date), Lundmark's correspondence, LUB.

[261] Knut Lundmark, "The Spiral Nebula Messier 33", *PASP* vol 33 (1921), 324-327.

[262] Lundmark, "Messier 33", 326.

[263] Knut Lundmark to Östen Bergstrand 16 December 1921, Bergstrand's collection, NC736, UUB.

[264] Harlow Shapley to Knut Lundmark 27 December 1921, Lundmark correspondence, LUB.

stated that van Maanen's measures "practically eliminate" the theory.[265] Writing to van Maanen, Shapley was clear that the measures of van Maanen were the ones with priority – arguments had to be harder than van Maanen's measures to make Shapley change his mind. He also stated that Lundmark was "lured by his own hypothesis" and that was why he failed to recognise the problem posed by the rapidly rotating nebulæ.[266] The next move from Lundmark came in print in April of 1922, where he gave more arguments for the nebulæ being distant objects. Shapley wrote van Maanen that Lundmark had sweeping generalisations and weak arguments; he hoped that van Maanen would enter the debate. Shapley would rather not come in to a public debate with Lundmark, as he saw Lundmark as sensitive. He asked van Maanen if people took Lundmark seriously.[267] He wrote Lundmark a long letter in which he listed several points where they did not agree. Shapley seemed seriously interested in avoiding a large public row with Lundmark, at the same time he wanted to present his arguments to Lundmark; he noted that none other than Henry Norris Russell was critical of Lundmark's paper.[268]

*

When Lundmark came to the Mount Wilson observatory, he was presented with a situation that was different than at Lick; there, Curtis had argued for the island universe hypothesis, and the observatory was more geared toward that cosmology.[269] At Mount Wilson van Maanen had measured rapid rotation in spirals – thereby presenting a strong argument for the local hypothesis – and Shapley had, before going to Harvard, worked out his large galaxy and critical stance towards the island universe. However, opinions towards the island universe theory were about to change at the Mount Wilson observatory.

---

[265] Harlow Shapley to Knut Lundmark 10 January 1922, Lundmark's correspondence, LUB.

[266] Harlow Shapley to Adrian van Maanen 14 January 1922, cited from Gale E. Christianson, *Edwin Hubble: Mariner of the Nebulæ* (New York, 1995), 156.

[267] Knut Lundmark, "On the Motions of Spirals", *PASP* vol 34 (1922), 108-115; letter from Shapley to van Maanen 19 June 1922 cited in Christianson, *Edwin Hubble*, 156.

[268] Harlow Shapley to Knut Lundmark 15 July 1922, Lundmark correspondence LUB.

[269] Robert W. Smith, "Cosmology 1900-1931", Hetherington ed., *Encyclopedia of Cosmology*, 116-126, 120.

Lundmark wrote a short paper together with Edwin Hubble on the nova that had appeared in the nebula NGC5253. They observed the spectrum of the spiral with the 10 inch Cooke telescope, finding it to be of the stellar type, like stars of about spectral type G.[270]

At Mount Wilson he also took up the very stumbling block of his cosmology: van Maanen's measurements of rotating nebulæ.[271] Several astronomers felt that the measures by van Maanen were very strong evidence against Lundmark and other proponents of the island universe model. Bertil Lindblad argued in 1920 that "it seems hard to get around this [van Maanen's] result, if not some extraordinary source of error exists"; two years later he stood "as a question mark between Scylla and Charybdis – the rotations and the spectral data [...] where M33, the Andromeda nebula etc. are to be placed you and van Maanen will have to fight it out."[272] Peter Doig wrote Lundmark in 1922 that van Maanen's work argued against the island universe hypothesis, and that the novæ seen in them probably were of another kind than the ones in our stellar system.[273] Years later, Shapley wrote Lundmark telling him that the reason he had sided against Lundmark's view of the cosmos was that he trusted van Maanen's measurements more than Lundmark's use of novæ as distance indicators.[274] van Maanen had a reputation for being a meticulous measurer of photographic plates.

At the core of the debate lay the relative merits of distance indicators. The astrometrical measurements by van Maanen belonged to classical as-

---

[270] Knut Lundmark and Edwin Hubble, "Nova Z Centauri", *PASP* vol. 34 (1922), 292-293.

[271] R. Berendzen and R. Hart, "Adriaan van Maanen's Influence on the Island Universe Theory", *JHA* vol 54 (1973), 46-56, 73-98; N.S. Hetherington, "Adriaan van Maanen on the Significance of Internal Motions in Spiral Nebulæ", *JHA* vol 5 (1974), 52-53; N.S. Hetherington, "The Simultaneous 'Discovery' of Internal Motions in Spiral Nebulæ", *JHA* vol 6 (1975), 115-145; N.S. Hetherington, *Science and Objectivity: Episodes in the History of Astronomy* (Iowa City, 1988); N.S. Hetherington & R.S. Brashear, "The Hubble-van Maanen Conflict Over Internal Motions in Spiralæ; yet More Information on an Already Old Topic", *Vistas in Astronomy* vol 34 (1991), 415-423. David Dewhirst & Michael Hoskin, "The Message of Starlight: The Rise of Astrophysics", in Michael Hoskin ed., *The Cambridge Illustrated History of Astronomy* (Cambridge, 1997), 256-343, 326ff.

[272] Bertil Lindblad to Lundmark 27 October 1920, 4 February 1922, Lundmark's correspondence, LUB.

[273] Peter Doig to Lundmark 26 October 1922, LUB.

[274] Harlow Shapley to Lundmark 12 January (could also be 1 December; the dating is rather unclear) 1930, Lundmark's correspondence, LUB.

tronomy with a long history of proven precision work; astrometry was a fundamental work method in astronomy. It did not involve any assumptions about the physical nature of objects far away in space, but it simply measured the positions of the nebular knots very accurately. The amount of motions and their systematic trends were then used to elaborate on the nature of nebulæ: since rotations were quite rapid, the objects could not be that far.

The nova method hinged on the assumption that the stars that flared up were of the same kind as novæ in our neighborhood, and that the luminosity of these nearby novæ could be determined to calibrate the method. The astrometrical work of van Maanen did not have that kind of opening to principal problems; but on the other hand van Maanen's work hinged on the measurement of very small motions in the nebulæ over a number of years, somewhere in the zone where the limit between signal and noise is drawn.

According to Lundmark it was W.S. Adams, the director at Mount Wilson, that had proposed that he should measure the plates of M33. At the Mount Wilson office in Pasadena he measured 400 points in the great nebula, before leaving back home for Sweden. The measurements were made on the same plates and with the same measuring engine as van Maanen.[275]

The rather time-consuming work of reducing the measurements was performed after Lundmark had returned to Uppsala in 1923. Raw measurements made at the measuring engine in Pasadena were reduced in Uppsala. Motions were far smaller than the ones measured by van Maanen. Lundmark's data gave a rotation period of 2.8 million years, while van Maanen's result was 230 000. Lundmark furthermore showed that there was a correlation between his and van Maanen's measurements.[276]

Before publication, Lundmark chose to make the results known in a letter circulated to amongst others Östen Bergstrand, Axel Corlin, Elis Strömgren, C.V.L. Charlier, van Rhijn, Schouten and Max Wolf. Bergstrand and Charlier wanted Lundmark to publish the results immedi-

---

[275] Knut Lundmark, "Studies of Anagalactic Nebulæ. First Paper", *Nova Acta RegiæSocietatis Scientiarum Upsaliensis.* Volumen Extra Ordinem Editum 1927 (Uppsala, 1927), 42.

[276] Knut Lundmark, "Internal Motions of Messier 33", *ApJ* vol 63 (1926), 67-71.

ately.[277] The result was viewed as rather controversial. Lundmark's teacher Bergstrand told him that there was a risk that the result would not be very well received. Some astronomers would probably be sceptical as to how objective Lundmark was, and would argue that he had forced measures in the way he wanted: "for sure, to begin with one will argue that you have chosen a certain standpoint and then have done everything possible to end up in that position".[278]

Lundmark had presented his result in the letter circulated in the summer of 1924 (it is mentioned in the letters for the first time in July); this is about one year after he left USA for Sweden. The long delay is explained by the many calculations needed to reduce the measurements (and also because of other duties). But another factor might be present, to explain why he issued the result at that particular time. In May, Gustaf Strömberg had written to Lundmark from Mt Wilson:

> Here everything is as usual and no changes have occurred. I can mention that Hubble has found a variable star in the Andromeda Nebula. If it is a cepheid and using Shapley's relation he finds the distance to be about 800 000 light years, a result that is liked by everyone here except van Maanen.[279]

It is not certain if this letter reached Lundmark before he finished the reduction of his measurements, but that is probably the case. Lundmark now had a rival in the discovery of the island universes, which perhaps prompted him to take publishing action. The measurements were published the year after in a paper dated June 1925.[280]

*

For some reason, Lundmark's work – the indications of the large distances of the nebulæ put forward by Lundmark capped with his development of the nova method and the 'refutation' of van Maanen's rotating nebulæ – did not make it all the way. Once considered the foremost promoter of island universes by Shapley, Lundmark witnessed Hubble

---

[277] Knut Lundmark to Elis Strömgren 17 July 1924, Strömgren papers, LUB; Axel Corlin to Knut Lundmark 12 August 1924, Lundmark's correspondence, LUB.

[278] Östen Bergstrand to Knut Lundmark 28 July 1924, Lundmark's correspondence, LUB.

[279] Gustaf Strömberg to Knut Lundmark 7 May 1924, Lundmark's correspondence, LUB.

[280] Knut Lundmark, "The Motions and the Distances of Spiral Nebulæ", *MNRAS* vol. 85 (1925), 865-894.

become the one who was seen as the discoverer of the distant nebulæ. In the mid-1930's, when the distribution of priority was becoming apparent, Lundmark lamented that there was a wave of "Hubbleianism" going through the astronomical world, and that his "sworn enemy [*arvfiende*] Hubble on Mount Wilson" got too much credit.[281]

One factor has to do with Lundmark's place in the geography of astronomy. He had as visiting scientist at the large telescopes a temporary possibility to make a few number of measurements with the large instruments, but not for long. As a visitor, he could not count on getting a good part of the observing time on the big telescopes, observing schedules had to be followed; he could also use the plate-stacks. Once on the mountains it was of course not at all sure that Lundmark would get much time on the telescopes. And after having spent some time there, he had to go home to Sweden with its humbler resources. Hubble, as one of the staff, of course had ample time to work with the largest instruments in the world. Therefore he had the possibility of using the large telescopes more continuously, giving him a better chance of making a discovery like the finding of the cepheids.

The fact that Hubble was connected to an institute that handled several of the largest telescopes in the world is also a factor in the way things turned out. The Mount Wilson observatory had become a centre for the work on nebulæ; a result made at such an institute thus could get more visibility because of the institutional home. An astronomer positioned in the astronomical periphery had a hard time making a good case. That was the way Lundmark saw the history when he looked back in early 1939: "during many years it seemed rather unlikely that I would succeed in laying my fair claims on priority as opposed to Hubble and his many supporters. It is always hard for a scientist from a lesser nation to make himself heard among the more exposed work of the greater nations."[282] To this shall be added the personality of Edwin Hubble: he was not one to subdue his own work.[283]

---

[281] Knut Lundmark to Harlow Shapley 5 April 1934, Knut Lundmark to Per Collinder 11 October 1934, Lundmark's correspondence, LUB.

[282] Knut Lundmark to Carl Bergsoe 15 January 1939, Lundmark's correspondence, LUB.

[283] Gale Christianson, *Edwin Hubble*.

Classification of Nebulæ

Early in his career, Knut Lundmark had begun by doing photographic observations of the colours of nebulæ with a method that had been deviced by his teacher Östen Bergstrand. In his PhD thesis he had mainly worked with material observed by other astronomers, whereas Bergstrand's other pupil Bertil Lindblad - Lundmark could not help but compare his career with Lindblad's - had continued more closely in the footsteps of Bergstrand and, eventually, Lindblad had developed the methods of Bergstrand farther. Lundmark's work was evolving away from the Uppsala way of doing astronomy.[284]

*

The photographic technology had strongly increased the number of nebulæ observed. Keeler had found large amounts of previously unknown faint objects, many of them spiral nebulæ, scattered over the sky on photographs with the Crossley reflector; he estimated the number of nebulæ within reach of that telescope to 120 000.[285] Max Wolf in Germany also photographed many nebulæ. A new territory of celestial objects had been opened up; the rapid increase of known nebulæ pressed forward the need for cataloguing and classifying these objects. Nebular astronomy needed a classification system.

When Lundmark spent some time at the Lick observatory he began using the rich plate material collected with the Crossley reflector. Going through the plate collection like a scholar using a library, or a botanist using a herbarium, Lundmark started a work of classifying nebulæ. The objects were classified according to a system devised by Wolf, but Lundmark added categories of his own. For about 400 objects he noted size, index of concentration (Lundmark's estimate of how concentrated the light was to the centre of the nebula), the brightness of the central condensation and the direction of the spiral arms.[286]

---

[284] Knut Lundmark to Östen Bergstrand 18 September & 23 September 1924, Bergstrand's collection NC736, UUB.

[285] Donald Osterbrock, *James E. Keeler: Pioneer American Astrophysicist and the Early Development of American Astrophysics* (Cambridge, 1984), 326.

[286] Pekka Teerikorpo, "Lundmark's unpublished 1922 nebula classification", *JHA* vol 20 (1989), 165-170, 167.

In 1925 he published a paper dealing with the problems of the nebulæ; it was an update on what had happened in nebular research since he wrote the PhD thesis five years earlier. In this paper, Lundmark put forward a proposal for a system of classification of nebulæ; he put the nebulæ in the order globular, elliptical, early spirals, late spirals, and nebulæ of the Magellanic Cloud type.[287] In 1925, the International Astronomical Union held a general assembly in Cambridge, where the subject of nebular classification was discussed. A system of classification was proposed by Hubble, but it did not win acceptance. Lundmark mentioned Hubble's system in a paper about the congress in Cambridge. Classification of nebulæ was seen by Lundmark as a very important field, but he argued that Hubble's system hinged on astrophysical observations. What was needed, instead, was a system of classification that was descriptive, based on photographically observed standards.[288]

In mid-October of 1925 the Swedish Astronomical Society convened in Lund. Lundmark gave a talk on classification systems. The classification of nebulæ ought to be done on the shape of the objects rather than measurements of the spectrum or colour index. He identified two broad groups, galactic nebulæ - belonging to our stellar system - and non-galactic, being outside of our system. Three broad groups constituted the non-galactic objects: elliptical, spiral and the magellanic type. The latter, chaotic, group had the fewest members.[289] Such a broad partition could hardly constitute a classification of the rich material collected. Lundmark soon produced such an extended system.[290] He argued here that it would be hard to classify the nebulæ, since there was such a wide variation of forms among the objects. But the task was important: a system of classification was needed for statistical analysis on the large numbers of observa-

[287] Knut Lundmark, "The Motions and the Distances of Spiral Nebulæ", *MNRAS* vol 85 (1925), 865-894, 867. (In this paper Lundmark also put forward a relation between the distances of nebulæ and their radial velocities, that later became famous as Hubble's law. Lundmark had a quadratic relation, Hubble a linear. Lundmark also concluded "that we have to deal with two distinct classes of Novæ". S Andromedæ and Tycho's nova were seen as examples of a class of novæ much brighter than ordinary novæ. Knut Lundmark, "The Motions and Distances", 887.)

[288] Knut Lundmark, "Den internationella astronomiska unionen", *PAT* vol 6 (1925), 169-173, 171.

[289] Knut Lundmark, "Om spiralnebulosornas indelning, dimensioner och utveckling", *PAT* vol 6 (1925), 104-114.

[290] Knut Lundmark, "A Preliminary Classification of Nebulæ", *Arkiv för matematik, astronomi och fysik* vol 9B (1926), no 8.

tions possible with the new generation of large telescopes. In addition, the possibility of outlining the evolutionary history of nebulæ was mentioned by Lundmark.[291]

Controversy between Hubble and Lundmark flared up. Soon after Lundmark's paper was published, Hubble accused Lundmark of plagiarism. Hubble had, without any success, tried to convince the IAU:s commission on nebular astronomy to accept his proposed system of classification. In the spring of 1926 Hubble had nevertheless decided to publish his proposed system, without the sanction of the IAU; before his paper had got in print, he learned of Lundmark's paper on the same subject.[292] A furious Hubble accused Lundmark of stealing his idea. Lundmark, recently elected to IAU:s commission on nebulæ and present at the conference in Cambridge, was accused of simply copying Hubble's work on classification, which was discussed at Cambridge.

Hubble prepared an attack, to be published in a central place in the astronomical literature, the *Astrophysical Journal*. Giving Lundmark an advance peek at was coming, he wrote a letter with a short quote from the paper, which he said contained

> a very mild expression of my personal opinion of your conduct and unless you can explain in some unexpected manner, I shall take considerable pleasure in calling constant and emphatic attention, whenever occasion is given, to your curious ideas of ethics. Can you suppose that colleagues will welcome your presence when they realize that it is necessary to publish before they discuss their work?[293]

When the paper was published, it did contain an attack on Lundmark that, considering the forum it was published in, was a direct accusation of plagiarism; if this was a mild expression of Hubble's views, he was really angry. "Lundmark, who was present at the Cambridge meeting [...] has recently published [...] a classification, which, except for nomenclature, is practically identical with that submitted by me. Dr. Lundmark makes no acknowledgements or references".[294]

Lundmark's reply came in "Studies of Anagalactic Nebulæ: First Paper", published in 1927 in the Acta of the Uppsala Scientific Society, a venerable scientific society with a history dating to the early eighteenth

---

[291] Knut Lundmark, "Preliminary Classification", 1.

[292] Christianson, *Hubble*, 175ff.

[293] Edwin Hubble to Knut Lundmark 26 August 1926, Lundmark's correspondence, LUB.

[294] Edwin Hubble, "Extragalactic Nebulæ", *ApJ* vol 64 (1926), 321-369, 323.

century, but its publication was not at all such a central place in the literature as the *Astrophysical Journal*. Hubble's attack in *ApJ*, Lundmark claimed, "is written in such a tone that I hesitate to give any answer at all".[295] Lundmark argued that he had been at the IAU general assembly in Cambridge, but did not at the time belong to the nebular commission; he did not then know of Hubble's proposed system of classification, which he learnt of through Hubble's letter in August 1926. He also argued that his system was not at all practically identical to that of his Californian adversary. Subclasses were classified by Hubble on the elongation of elliptical objects or the form of spiral arms, while Lundmark used the level of concentration of light towards the centre of the object. He also argued that two of the three main groups of nebulæ – elliptical, spiral – had been in use since the mid-1800's, when they were discussed by Lord Rosse and Alexander; the importance of the third, irregular, form had been pointed out by Lundmark in a paper published well in advance of Hubble's work, in 1924.[296]

Lundmark had indirect support from another central astronomer who also was working in the classification of nebular objects: Harlow Shapley. Shapley, too, was critical of Hubble's system and published a proposal of a system. But unlike Hubble's confrontational style, Shapley was open towards Lundmark. He wrote that he hoped to discuss further cooperation on the classification of nebulæ.[297]

Hubble's accusation of plagiarism has been questioned by Pekka Teerikorpi, who has found an unpublished classification of nebulæ made by Lundmark during his time at Lick in 1922. He also cites a letter from 1922 in which Lundmark discusses classification of nebulæ.[298] It is not unusual that cases of nearly simultaneous work lead to disputes of priority; this seems to be another example of that. Hubble's insisting on his priority also fits with the personality of Hubble as it is depicted in Gale

[295] Knut Lundmark, "Studies of Anagalactic Nebulæ: First Paper", *Nova Acta Regiæ Societatis Scientiarum Upsaliensis* volumen extra ordinem editum 1927 (Uppsala, 1927).

[296] The discussion of the irregular type is Knut Lundmark, "The Distance of the Large Magellanic Cloud", *Observatory* vol 47 (1924), 276-279, 277.

[297] Harlow Shapley, "On the Classification of Extra-Galactic Nebulæ", *Harvard College Observatory Bulletin* no 849 (1927); Harlow Shapley to Knut Lundmark 3 July 1926, Lundmark's correspondence, LUB.

[298] Pekka Teerikorpi, "Lundmark's Unpublished".

Christianson's biography. Hubble was a very self-centred climber (of both social and scientific ladders).[299]

Lundmark had managed to become on very unfriendly terms with a person that soon was to be considered almost a modern Copernicus, a very central person in the way the story of the important discoveries in observational cosmology in the 1920's came to be presented.

Hubble's attempts to put his own work in the foreground were very successful. In the simple but effective historiography that lives amongst many astronomers and in many popularisations of astronomy, Hubble is often portrayed as one of the few heroes of twentieth century astronomy. The episode with Lundmark hints at the simplicity of such a picture.

The Mobility of Data and the Lund General Catalogue

When Lundmark came to California in 1921 he had the chance to use the large instruments there for short periods of time. Of more importance to him were the large stacks of photographic plates: archives of the data observed with large telescopes. This kind of material was a very important component in the work Lundmark tried to stage when he came back to Sweden. The work on classification of nebulæ was something that could be done without telescopes. Lundmark left California to a position quite distant from rich collections in plate vaults and good telescopes under clear skies; in various ways he tried to bridge that distance. He worked with material from different parts of the world and had as an overarching goal the ambition to put these pieces of heterogeneous data together in a homogeneous system. He had to devise strategies and tactics that enabled empirical material to be collected and organised. If successful, it would allow him to have a strong position in nebular astronomy, although he would be working out of Sweden, with scant resources in these areas.

Part of the time in California had been spent working out a catalogue of nebulæ from material in the collections of plates at Lick. Extracting

[299] A parallel case was when Willem de Sitter in 1931 hinted that Hubble might not have sole priority in the discovery of what became known as Hubble's law; Hubble's reaction was strong, and de Sitter withdrew. Christianson, *Hubble*, 230.

data from the plates stored at Lick was a way of improving the mobility of the data, to make it possible to bring some of the data back home.[300]

Lundmark tried various ways to get to material that could further his work on nebulæ. One was borrowing plates. Upon leaving Lick, he had some plates with him, including several of the large nebula M33 in the constellation Triangulum.[301] In early 1926 he got a letter reminding him that the plates ought to be returned; at Lick, there was not one good plate present of M33, since Lundmark had borrowed the material. "We cannot reasonably permit the negatives to be much longer away from their home", William Wallace Campbell wrote.[302] A year later, Lundmark still had the plates at Uppsala. A letter in a more harsh tone came from Robert Grant Aitken, Campbell's successor as director of Lick:

> It is now fully four years since you borrowed certain Crossley negatives of nebulæ, including M33, and that would seem to be an interval of time sufficiently long to enable you to get out of the plates any data that may be of interest to you. In the meantime, we have suffered the inconvenience and really the hardship of not having available for our own use here our own original negatives of these objects! I think that, upon consideration, you will realize that it is now your duty, as one who has enjoyed our hospitality and received favors from us, to return all Lick Observatory negatives in your possession without delay, packing them securely in wooden boxes, with plenty of shock-absorbing material to assure safe transportation and sending them sealed and registered. [...] We were glad to put this material at your service, but in doing so we had no expectation that it would be retained for a period longer than a year, or at the very longest, two years. I think that if I were to make the facts known to any astronomer, either in your country or mine, or to the Rector of the University of Upsala, he would agree that we have been exceedingly patient in waiting upon you so long.[303]

Working out strategies and tactics to enable access to data was a major task for him, and sometimes he ran into considerable trouble in doing so.

---

[300] After his return to Sweden from the US, Lundmark also worked during the summers of 1924 and 1925 at Greenwich. There he used the Franklin Adams charts, a collection of wide-field photographs that covered the whole sky. With P.J. Melotte he studied the distribution of dark nebulæ, and produced a catalogue of these objects. Anita Sundman, *Den befriade himlen*, 59ff.

[301] Robert Grant Aitken to Knut Lundmark 26 March, 31 March 1924, Lundmark's correspondence, LUB.

[302] William Wallace Campbell to Lundmark 8 January 1926, Lundmark's correspondence, LUB.

[303] Robert Grant Aitken to Lundmark 21 February 1927, Lundmark's correspondence LUB. (Aitken's emphasis.)

This process in some ways explains his career. Lundmark had to device ways of countering his placement in what in some ways could be considered the scientific periphery: in Sweden the observatories were not suitable for doing nebular astronomy and the skies are often overcast. He tried to gain access to data in more central parts of the scientific geography, and in the end construct a scientific data centre, to have data flowing to him.

*

Lundmark succeeded Charlier as the chair of astronomy in Lund in early 1929 (see chapter six below). After becoming professor, Lundmark pursued a project of classification and cataloguing; this programme went back to cataloguing work he had done prior to coming to Lund.[304] One background to this project was Lundmark's own placement. These activities could be pursued without immediate access to large telescopes in a favourable climate. He tried to build a network to enable the flow of scientific data to his rather peripheral position; he tried to make Lund a centre of calculation and data treatment.

A short catalogue produced by Lundmark in the early 1920's developed into a larger project towards the end of the decade. In late 1927 the project had reached such a stage that he began to apply for money to it. He tried to persuade the Swedish Astronomical Society apply for 5 300 kr from the state for such a project. The project application gives insights in how Lundmark argued for the kind of astronomy he was pursuing.[305] He described here how the nebular research had taken a step forward. The advances during the latest years on the determinations of distances to the objects among other things had shown them to be anagalactic systems (Lundmark's term for stellar systems outside of the Milky Way system). Other areas of nebular astronomy were however about to be opened up, Lundmark argued. While the structures, dimensions and luminosities of nebulæ differed, the spread in subgroups (defined "on the basis of a study

---

[304] For a short introduction to the history of nebular catalogues, see Peter Nilson, "Catalogues of Galaxies. A Short Survey of their History", in Peter Nilson, *Uppsala General Catalogue of Galaxies* (*Nova Acta Regiæ Societatis Scientiarum Upsaliensis* ser V:A vol 1 = *Uppsala observatoriums annaler* vol 6) (Uppsala, 1973); Peter Nilson, "Knut Lundmark and 'Lund General Catalogue of Nebulæ'", unpublished ms., copy in the author's collection.

[305] Knut Lundmark, "Till Svenska Astronomiska Sällskapet", December 28 1927, in Peter Nilson's private collection, copy in the author's collection.

of the outer structure") was definitely smaller. Studies of these subgroups of nebulæ could lead to advances in theories of the "genetical relation" between nebulæ. One way of reaching the goals would be a photographic survey of all the sky. This had been discussed in the International Astronomical Union and was viewed as impossible: it would require an excessive amount of observations, because large telescopes had small fields of view. Also, "every attempt at uniformity in the material must be abandoned", since photographs had to be exposed with telescopes of different kinds. Other types of instruments with larger fields of view could perhaps be employed, but the development of such, Lundmark argued, lay in the future.

In the grant application, Lundmark claimed that he himself could not count on being part of photographic survey work:

> However the issue of a photographic survey of the nebulæ will be solved, it is clear that the realization of such a project will under all circumstances take many years, and definite results are surely not to be expected until decades later. For my own part I cannot count on taking part more substantially in such a survey, since my activity during the foreseeable future will be based at Swedish observatories with instrumental resources that are not fit for such work.[306]

Lundmark had identified the need for a photographic survey of nebulæ, the enormous size of such an undertaking with the telescope technology of his time, and the insuitability of the instruments at Swedish observatories for such a project. He had a way of moving around these difficulties. He devised a project that aimed at collecting and statistically discuss "all the existing information of different kinds on the anagalactic objects".[307]

This would give him a task where he could work on what he perceived as modern front line research on the nebulæ without the needs for the telescope resources of a large and modern observatory. The material would come from different sources: publications that specifically dealt with nebulæ; from the snippets of information about nebulæ, sporadically published "all over the astronomical literature" (his project is of a very large scale – he envisions a survey of all the astronomical literature); from unpublished material that Lundmark himself had produced, as well as material that other astronomers would contribute to the project.

[306] Ibid.
[307] Ibid.

Observations from different astronomers in different times (hence, made with a variety of types of instruments) would be reduced to a common system, whereupon statistical studies could be made on the corpus of data. Data from widely differing observers would be pieced together to a whole; differences in observational techniques should be done away with in a process of homogenisation. Nebular data should become comparable and combinable, a systematic dataset that could be used by other astronomers. Lundmark saw the *Lund General Catalogue* as a "far-reaching process of statistical homogenisation, with the goal of welding together observations made by different astronomers into one series of observations, as if they were made by a fictitious average observer."[308] It became Lundmark's task to construct such a fictitious observer. He even had the ambition of tying together the work of modern photographic era with the data from the visual age: the "observations by William and John Herschel represents an invaluable material, when it comes to enlarging the present knowledge of the gigantic world system of the nebulæ."[309]

Twelve thousand index cards were to be the physical format of the data, and the funding he applied for was for the assistant needed to enter data on the cards and do simple calculations. In July of 1929 2 500 Swedish crowns were awarded from the state to Lundmark via the Swedish Astronomical Association.[310] Resources for the project also came from the International Astronomical Union; $150 was allotted, with the help of Harlow Shapley and Frank Schlesinger.[311] The resources Lundmark obtained for the project thus were thin, compared to several other observatories. Still, he had the use of the crew of assistants, a legacy from Charlier's time as professor.

In the yearly report of the observatory for the academic year 1929-1930, the first Lundmark authored as professor, he described a visit to American observatories between March and December of 1929. He used the plate-collections (rather than the telescopes) at Mount Wilson, Lick and Lowell observatories. He measured and classified about 8 000 nebulæ,

---

[308] Yearly report of the Lund observatory, *LUÅ* 1931-1932, 33-46, 36.

[309] Knut Lundmark, "Till Svenska Astronomiska Sällskapet".

[310] "Avskrift av Kungl. Maj:ts nådiga brev till statskontoret", copy to the Swedish Astronomical Association May 16 1929, in the collection of Peter Nilson, copy in the author's collection.

[311] Harlow Shapley to Knut Lundmark August 16 1928, Lundmark's correspondence, LUB.

data that was to be entered in to the general catalogue of nebulæ.[312] In a short paper written in the summer of 1929, when working at Mount Wilson, he described the plans of what he called The General Catalogue of Nebulæ of the Observatory of Lund. The plans called for a catalogue of nothing less than "all existing data" on nebulæ, entered on cards of the size 20 by 30 centimetres, with data on both sides. By the summer of 1929, data had been entered on 16 000 cards; he thought the total catalogue would reach 35 000 cards. The catalogue was in 1929 placed at Uppsala, where his assistant and niece Ingeborg Lundmark worked on the routine parts of the project. It was, however, to be transported to the observatory at Lund, where future data would continuously be entered. Data should not only flow in to Lund; after it had been homogenised, standardised and entered in the catalogue, data would flow out of Lund. The Lund observatory was to serve the astronomical community with data from the catalogue gratis. Lundmark had a vision of the Lund observatory as a data centre in the world of nebular astronomy. In the future, the catalogue would also be published.[313]

Lundmark and his staff at the observatory moved forward quickly. In mid-1931, 30 000 objects had been entered in the catalogue, whose total volume then was seen as 40 000 objects.[314] The astronomical literature was combed for information, Lundmark's visits to observatories resulted in new data to be entered, and Lundmark's pupils often had instructions to take data with them when returning home. Several pupils were sent out on such data collecting missions to other observatories.

Lundmark visited the United States again in 1932 and once again got access to plate archives. "I doubt, however, the possibility of offering you much time with any of the telescopes", Robert Grant Aitken wrote Lundmark. Arrangements were later made for Lundmark to have copies made of plates in the Lick collection.[315]

At Steward observatory in Arizona, the 1 metre reflector was suitable for photographing nebulæ. Lundmark's graduate student Sture Holm was there for some time, and acted as Lundmark's agent. A list of interesting

---

[312] Knut Lundmark, yearly report of the Lund observatory, *LUÅ* 1929-1930, 29-35, 29f.

[313] Knut Lundmark, "A New General Catalogue of Nebulæ", *PASP* vol 42 (1930), 31-33.

[314] Yearly report of the Lund observatory, *LUÅ* 1931, 36-46, 39.

[315] Robert G. Aitken to Knut Lundmark 4 April 1933; W.H. Wright to Knut Lundmark 4 November 1938, Lundmark's correspondence, LUB.

double nebulæ was provided from Lund for future observation at Steward observatory.[316]

A telescope of similar size was available at Heidelberg. Two of Lundmark's graduate students went there. The first, Erik Holmberg, worked at Heidelberg in 1936. "Do whatever you can, both to try to get me as many reproductions as possible and to expand the particularly meagre collection of data with new matters of empirical fact" Lundmark wrote Holmberg. The year after, Anders Reiz was at Heidelberg, where he had rather free access to the plate archives; among other things, he was on the lookout for supernovæ on the Heidelberg plates.[317]

Lundmark had contacts with astronomers in Åbo. Here a program for photographic studies of asteroids was in operation; Lundmark used these plates also for studying nebulæ.[318] In Egypt, the Helwan observatory also had a quite large reflector. Lundmark's graduate student Björn Svenonius worked there for a while and Lundmark asked Svenonius to "try to borrow as many plates of the large spiral nebulæ around the southern galactic pole as possible. I am especially thinking of NGC 55, 134, 300, 1097 and several others that show resolved stars." Lundmark wanted the plates for his own work, and asked Svenonius to make a hint to the observatory director Madwar about honorary doctorates, medals and other possible rewards that could be given to the observatory director who opened his plate cabinets. The hint was successful. Madwar had in the beginning, Svenonius wrote, been rather negative to lending plates ("even plates that I had exposed and paid for myself"), but when Svenonius mentioned "that it might open doors to a Swedish honorary degree for him, he changed completely and is now willing to lend you Helwan's complete plate library, if you would want to".[319]

Lundmark thus got access to empirical data observed at other observatories, by visiting these institutes where he analysed material, having cop-

---

316 Edwin F. Carpenter to Knut Lundmark 7 May 1932, 3 August 1934; Sture Holm to Lundmark 24 January 1933, Lundmark's correspondence, LUB.

317 Knut Lundmark to Erik Holmberg (copy) 24 September 1936. In corresponding with Lundmark, Holmberg wrote among other things of how the whole observatory staff went to Karlsruhe to hear Hitler give a speech. Erik Holmberg to Knut Lundmark, no date. Anders Reiz to Knut Lundmark 14 November 14, 22 November 1937. All in Lundmark's correspondence, LUB.

318 Inez Masloff, "Knut Lundmark i Åbo", in Johnson ed., *Knut Lundmark*.

319 Knut Lundmark to Björn Svenonius 28 March 1935, Svenonius to Knut Lundmark 12 March 1935 and 6 April 1935 (the quote), Lundmark's correspondence, LUB.

ies made of the plates, or by using published observations. He could not direct telescopes: his project was piggybacked on other programmes; he had to find a way to use material collected for other purposes. His programme was a large and pragmatic accommodation to the resistance presented to Lundmark, a way to overcome his peripheral position.

Lundmark was a collector of scientific data, whose style of doing astronomy was similar to natural history. Perhaps it is not surprising that he also pursued botany as a hobby, and was an ardent book and stamp collector, and thought highly of Aristotle. Lundmark was searching for wholeness; he had an urge to piece together the little that was known about very many nebulæ into a homogeneous system. His way of doing astronomy hinged on thorough knowledge of the astronomical literature.

Lundmark's ambitious project of cataloguing the total knowledge on nebulæ never really was finished. It was worked upon for quite some time by Lundmark and his staff. The catalogue never made the Lund observatory the central repository of knowledge about nebulæ that Lundmark wanted. Because of its sheer size, Lundmark hoped the Lund General Catalogue would make Lund a central place in nebular astronomy. Also because of its sheer size, the project never reached completion, or for that matter was ever used by international astronomers. Data was produced by telescopes faster than Lundmark could adapt it to his catalogue. Also, he did not make use of the automatic sorting that became available at the time. Punched card machines had begun to be used; among others, Frans Josua Linders employed such Hollerith machines at the eugenics institute at Uppsala. Such a machine would perhaps have increased the chances of success.

Lundmark's project seemed increasingly outdated. Other classification and cataloguing projects were being produced. In 1932, Harlow Shapley and Adelaide Ames published a catalogue of nebulæ that was much lesser in quantity than Lundmark's, but on the other hand was based on a much more homogeneous material.[320] Shapley and Ames used the plate stacks observed at Harvard observatory, a material that was far more homogeneous than the heterogeneous observations Lundmark tried to align into the Lund General Catalogue. Shapley and Ames tabulated positions for the epoch 1950 (Lundmark used the older standard 1900), diameters and classifications in Hubble's system of classification. The mate-

[320] Harlow Shapley and Adelaide Ames, *A Survey of the External Galaxies Brighter than the Thirteenth Magnitude* (*Annals of the Harvard College Observatory* vol 88, no 2) (Cambridge, Mass., 1932).

rial was much smaller than LGC but quite complete down to the limiting magnitude of 13. Astronomers doing statistics could be rather confident that most objects down to the limiting magnitude had been entered into it, and also that the data was uniform, since the classifications in it had been done by the same persons. This high quality catalogue put Lundmark's project in a different light. Had he published the results of his work up to about 1930, LGC might have made a difference. In the spring of 1931 about 30 000 objects had been entered into the catalogue, and a publication of this dataset might have made an impact on nebular astronomy. Lundmark tried to have the LGC embrace the Shapley-Ames catalogue by entering data from it into the LGC, but LGC never really made it in nebular astronomy.[321]

While Lundmark's project wasn't institutionalised internationally, his project nevertheless was important for the work at the Lund observatory. His co-workers and pupils studied problems in nebular astronomy, led to such subjects by Lundmark's interest in those fields. Several of the PhD thesis projects made under Lundmark's time can be seen as being projects that extended Lundmark's vision of nebular astronomy. These projects contained catalogues that can be seen as parts of the LGC.

The grand project never reached completion. However, it provided an umbrella of sorts under which the astronomers working at the observatory could develop their careers.[322] Several PhD projects were started on subjects that dealt with issues close to the LGC project: morphology of galaxies, cataloguing of galaxies and clusters. Although the LGC never made the Lund observatory into a centre for nebular astronomy on an international scale, some of the working methods of that project and the data collected materialised in the work by people under Lundmark, where a group of astronomers emerged that worked on nebulæ and galaxies. They used material brought to the Lund observatory on photographs, either as original plates or copies. In some cases, they used the

---

[321] Another and later reason why the LGC never caught on was the development of telescopes that combined two qualities that up until the mid-1930's had not been present in the same telescope: large aperture (so as to reach faint objects) and wide field (to reach many objects and a large part of the sky on a single photographic plate). Telescopes of the Schmidt type could photograph large areas of the sky, while reaching faint objects and also having good resolution. An early and large Schmidt camera was used at Caltech from 1936. Donald Osterbrock, "Getting the Picture: Wide-Field Astronomical Photography from Barnard to the Achromatic Schmidt, 1888-1992", *JHA* vol 25 (1994), 1-14.

[322] Peter Nilson, "Knut Lundmark".

plate stacks at foreign observatories. Working almost like the humanistic scholar, they travelled to plate libraries where they studied the collections, sometimes on copies brought home to Lund.

Erik Holmberg studied planetary companions around other stars. In a discussion of observations of trigonometric parallaxes, Holmberg noted that the stars often showed periodic displacements, effects probably due to gravitational perturbations by invisible companions to these stars. In some cases the perturbations were so small, as to lead Holmberg to conclude that they were produced by objects with masses in the Jupiter class, i.e. planets.[323] Holmberg's main work lay, however, in the study of nebulæ. In his PhD thesis published in 1937 he studied a class of galaxies, the double objects that had attracted Lundmark's interest before.[324] Lundmark had done work on the double galaxies and had found about 200 double and multiple systems.

Erik Holmberg's thesis used empirical material observed at the Heidelberg observatory at Königstuhl. Here, a large collection of plates was stored, the observational legacy of Max Wolf, who had used a photographic telescope with 0.4 meter aperture. Holmberg spent in total about six months during 1935 and 1936 in Heidelberg, searching for double galaxies on the plates. Holmberg had access to a plate material that covered 80% of the sky down to faint objects: the limiting magnitude was around magnitude 16, in some cases as faint as 17.5. From this material a catalogue of the properties of 827 double and multiple galaxies was produced. He also concluded that the formation of double galaxies probably was due to galaxies capturing other galaxies.

Holmberg continued to make a number of studies in nebular astronomy, during and after the war. He worked on the calibration of photometry of nebulæ and studied nebulæ photographically with the large reflectors on Mount Wilson during 1940 and 1941, and measured the plates photometrically (using a photometer at the Uppsala observatory).[325]

---

[323] Erik Holmberg, *Invisible Companions of Parallax Stars Revealed by Means of Modern Trigonometric Parallax Observations* (*Lund Medd.* ser II no 92) (Lund, 1938).

[324] Erik Holmberg, *A Study of Double and Multiple Galaxies: Together with Inquiries into some General Metagalactic Problems* (*Annals of the Observatory of Lund*, no 6) (Lund, 1937).

[325] Erik Holmberg, *On the Existence of a Systematic Error in the Estimated Magnitudes of Galaxies* (*Lund Medd.* ser II no 100) (Lund, 1939); Erik Holmberg, *A Photographic Photometry of the Spiral Nebulæ NGC 891 and NGC 3623* (*Lund Medd.* ser II no 114) (Lund, 1945).

He studied the absorption of light in nebulæ, and made experimental studies of errors in measurements of the sizes of nebular objects by measuring photographs of artificial nebulæ.[326]

Several other astronomers at Lund, like Björn Svenonius and Carl-Gustav Danver did nebular astronomy. They worked on the morphology and classification of galaxies by using photographs taken at other observatories.[327]

Photography was an important element in the handling of astronomical data at the Lund observatory. A key person was the observatory's photographer Gösta Hultzén, who was a skilled photographic technician. The methods devised by Lundmark and his pupils had as one cornerstone the transportation of scientific data from other astronomical observatories, using the photographic plates. Hence, photographic technologies were something that the Lund staff put some effort into developing.

The astronomers worked on various methods for enhancing the photographic material. For an astronomer studying the regions close to the bright nucleus, overexposure usually was a problem; the plate became blackened in the nuclear regions, thereby washing out any structure. The faint spiral arms needed a longer exposure to show up well. Astronomers with access to an observatory with proper instrumentation could take a series of plates with various exposure times, showing details from the nucleus out to the spiral arms.

The Lund astronomers often worked with plates already exposed; they could not determine the various exposure times of the plates. They therefore worked on methods of treating exposed plates that worked almost like they had had the possibility of observing with various exposure times. Faint images on the photographic plates were brought forward by repeated copying on high contrast plates. Structures in the bright nuclear regions were studied by reducing the amount of exposed silver on the

---

[326] Erik Holmberg, *On the Apparent Diameters and the Orientation in Space of Extragalactic Nebulæ* (*Lund Medd.* ser II no 117) (Lund, 1946); Erik Holmberg, *On the Absorption in the Spiral Nebulæ* (*Lund Medd.* ser II no 120) (Lund, 1947).

[327] Björn Svenonius, *Contributions to the Photographic Study of Colour, Resolution, and Total Magnitude of Anagalactic Objects with Special Regard to Objects Photographed at Helwan* (*Annals of the Observatory of Lund* no 7) (Lund, 1938). Carl-Gustav Danver, *A Morphological Investigation of Some Near Galaxies with Regard to the Length and the Form of their Arms, their Inclinations and their Symmetry Properties* (*Annals of the Observatory of Lund*, 10) (Lund, 1942).

plates.[328] Thus, the photographs taken once and for all at a remote observatory were being mined for information on the astronomical objects. The Lund astronomers seldom had access to large instruments during long periods. (That would have enabled them to determine themselves the features of the plate material: short exposures for studying nuclear regions of galaxies, longer exposures for following the faint parts of the spiral arms.) Instead, they had to devise photographic methods of getting that information out of the photographic plates.

The observatory arranged courses in scientific photography during the 1930's. While some courses were given on a temporary grant, Lundmark tried to use the photographic knowledge at the observatory to get more permanent funding for such courses.[329] At the international photographic exhibition in Stockholm in 1934, the Lund observatory participated in the scientific class with focus on the work at Lund. Enlargements made at Lund of plates exposed at Mount Wilson were shown.[330]

The photographic skills and routines developed at Lund were also put to use in the teaching of astronomy as well as in the popularisations of astronomy, pursued by Lundmark and his staff. Before Lundmark's entry as professor, the Lund observatory had made a collection of slides that were used in teaching.[331] This collection was expanded during Lundmark's time in Lund and was put to use when Lundmark started doing lecture tours of talks on popular science; they were also loaned to schools. Hultzén's skills were also very useful for Lundmark's publications in popular astronomy in magazines, as well as for Lundmark's wish to collect classical works from the history of astronomy, photocopies of which are still in the Lund observatory library.[332]

---

[328] Cf. Carl-Gustaf Danver, *Morphological Investigation*, 24-38; Björn Svenonius, *Photographic Study*, 42-69. At about the same time, Bertil Lindblad studied similar galactic features at the Stockholm observatory. He could, however, observe the structure of the galaxy NGC 7331 by exposing plates with the optimum exposure time using a telescope at his home institute. The example illustrates the different strategies developed at Lund and Saltsjöbaden for getting empirical material. Bertil Lindblad, *On the Distribution of Light-Intensity and Colour in the Spiral Nebula NGC 7331* (*Stock. Ann.*) vol 13 no 8 (1941).

[329] Knut Lundmark, "Kurser i vetenskaplig fotografi vid Lunds universitet", *Nordisk tidskrift för fotografi*, 1934 no 6, 77-80.

[330] *Internationella fotografiutställningen januari 1934. Vetenskapliga avdelningen* Liljevalchs konsthall, catalogue no 102, 45-53.

[331] *Katalog över diapositiv förfärdigade vid Lunds observatorim* (Lund, 1929).

[332] Knut Lundmark to Elsa Nyblom March 1 1935, Lundmark's correspondence, LUB.

The Open Observatory: Rhetoric and the Public

Lundmark's project of classifying and cataloguing nebulæ lost impetus during the 1930's. In several of the yearly reports in the second half of the decade, the project is not mentioned at all. In the heterogeneous collection of data that Lundmark tried working on, the literature was in a way the only easily accessible data. The old observatory with its outdated instruments could not be used to input data into the project; at Lund, observational work was almost impossible to pursue, but he still hoped he would eventually have access to a modern observatory, Lundmark wrote to the American astronomer S.A. Mitchell in November 1937.[333] Attempts were made to collect data by sending Lundmark or his students to foreign observatories, but this was not enough. The network in which the project operated never became strong enough.

Around 1935 Lundmark's activities changed. The project of classification and cataloguing had been going for a number of years and was grinding to a halt. Lundmark began devoting more and more of his time to popularisations of astronomy. He had been interested in popularisations from early on, and now he developed that part of his scientific style. The scientific work at the institute was shown to the public. Lundmark began a series of public nights at the observatory. Beginning the 9th of December 1935, the observatory was open every second Monday.[334] An accompanying booklet was published, which told the story of the observatory and introduced the present-day activities.[335] The number of visitors was large. From late 1935 until the end of 1939, the observatory had almost 3 500 visitors in total, and in the autumn semester of 1940, 1 270 visitors came to the observatory. Even today, one can hear the observatory's personnel speak about when Knut Lundmark addressed a crowd in the observatory gardens from loudspeakers. Lundmark's idea was a success, but after the spring semester of 1941, the open nights had to be cancelled be-

333 Knut Lundmark to S.A. Mitchell November 11 1937, Lundmark's correspondence, LUB.

334 Yearly report of the Lund observatory, *LUÅ* 1935-1936, 37f.

335 Knut Lundmark, *Astronomien i Lund 1667-1936: Glimtar och drag ur astronomiska institutionens liv och öden i nuvarande och gången tid* (Lund, 1937). Yearly reports of the Lund observatory, *LUÅ* 1936-1937, 40-43, 41.

cause it put too much a strain on the resources of the institution; thereafter, only groups from schools, societies, and the like were admitted.[336]

At the open nights a number of activities were available to the public. Visitors went to various stations. Lectures were held on some astronomical topic, illustrated with pictures from the rich collection of slides. In the lecture hall, enlargements of astronomical pictures were displayed. They saw how astronomers used the meridian circle to observe the positions of stars; they saw the astrograph in the observatory garden; they went down the stairs in the seismograph cellar. Visitors could see the computing bureau with the stacks of cards containing data (LGC and other datasets), calculating machines, the workplaces of the computers. They could also see the crammed spaces of an observatory that needed larger buildings. To many visitors, the highlight must have been the visit up in the tower of the old building, were the refractor was looking out at the sky.

Not only the wish to educate the public was the reason behind the open nights at the observatory. There was a tactic behind the activity: by opening up the institute, the problems facing the Lund astronomers would become apparent. Hopefully, sooner or later a movement would form among people with power to give the astronomers better resources. In the booklet that came with the open nights, Lundmark hinted at the remarkable expansion the university and the observatory were part of, and made remarks about the old instruments and the state of the buildings:

> The visitor can easily see that the buildings of the observatory are far too small even for the original purposes [...] The 'large' refractor, which we now stand in front of, is in fact a 'small' refractor or, in other words, with today's standards, quite modest [...] Because the number of people working at the observatory has been rising steadily, and collections and auxiliary equipment also have expanded, the buildings now are quite insufficient. The observatory also does important work in pure geophysics, not only with the meteorological work but also with observations of earthquakes and the penetrating cosmic radiation. [...] When we do get a new observatory, it is clear that a much bigger refractor will need to be installed.[337]

[336] Yearly report of the Lund observatory, *LUÅ* 1939-1940, 45; yearly report of the Lund observatory, *LUÅ* 1940-1941, 44; Bo Nilsson, private communication, August 1997.

[337] Knut Lundmark, *Astronomien i Lund*, 6, 20.

These are not the words of a proud institute leader who shows his department. Lundmark wanted to use the popularisation of astronomy to gain more resources for the observatory.

The refractor was an important instrument in the popularisation of astronomy at the Lund observatory. If the instrument was quite unusable as a tool for observational astronomy in the 1930's, it had all the look and feel of a telescope, and its size was more or less perfect for open nights at the observatory. After having toured the observatory and the various activities pursued there, visitors went up the stairs to the 'large refractor'. Five metres long and made out of polished wood and brass, it must have been an imposing sight to the visitor. The aperture, 25 centimetres, is large enough to allow detailed views of the moon and planets.

The refractor also shows up in articles in popular magazines, when Lundmark poses for the photographer.[338] In popular activities, the telescope found a use as a visual signal of astronomy. Scientists were in this time often portrayed with their instruments. The scientific instrument was a visual cue that signalled the presence of a scientist with his activity.[339]

Lundmark led an observatory with ageing instruments. This resource could be aligned with the ambition of doing popular astronomy, an activity for which the refractor was suited. Scientific instruments sometimes are flexible and plastic resources. They can be put to various use by the scientists, and they support changing agendas. One instrument can play new roles in various settings of scientific practice throughout long periods of time. In this process the instrument can change from having a centre stage to being decentred, when new instrumentation is brought to work with it in a mosaic fashion, to a situation where the instrument is dismantled, but parts of it still lives on in new alignments of scientific hardware. The case of the Lund refractor illustrates such a plasticity in

---

[338] Cf *Veckojournalen* no 10 1935. Writing to Elsa Nyblom, an editor at *Veckojournalen*, Lundmark told about how "our very competent photographer (Hultzén) did a very good work in helping with the most difficult instrument photographies. Sometimes we had the photographer hanging in dangerous positions in the windows of our tower and there were a lot of problems with lighting and so on. Everything went well, we worked bravely for hours and everything looked almost like a Hollywood set. I posed for up to 45 seconds for a single exposure and it is not easy to remain still for so long a time." Knut Lundmark to Elsa Nyblom March 1 1935, Lundmark's correspondence, LUB.

[339] Cf Sven Widmalm, "Porträtt av vetenskapsmän", Gunnar Broberg ed., *Gyllene äpplen: Svensk idéhistorisk läsebok* (Stockholm, 1991), 840-842.

scientific instruments. As shown above, Nils Dunér could momentarily transpose the instrument by adding auxiliary instrumentation that, together with the refractor, allowed him to do astrophysics in an institutional environment originally geared to classic astronomy. This process of change in the place of the refractor in scientific practice continued long after the period covered here.[340]

Scientific instruments can also be an interesting source for discussing the history of popularisations of science. Publications and talks are perhaps the major vehicles for doing popular science, but the use of scientific instruments are also resources for popular astronomy and could be used more by historians interested in discussing popularisation of science.

*

Lundmark was convinced that science belonged to the wider culture just as subjects like theology and history. For him, astronomy was not separated from culture; he was interested in the humanities, and claimed that astronomy also could give something to the humanities. It was, in his case, not obvious where the boundaries between humanities and science should be drawn. He frequently crossed the dividing line between science proper and popular culture; he spent a lot of time on popularisations of astronomy in print and via extensive lecturing across Sweden, as has been documented by Anita Sundman in *Den befriade himlen*.

Extraterrestrial life – that most popular of astronomical subjects – was also on Lundmark's agenda. In 1926 Lundmark published a book on life in the universe.[341] A more extensive work followed in 1935, *Livets välde*,

---

[340] After Lundmark's time the instrument was dismantled. The observatory no longer showed the sky to visitors; instead a planetarium was installed to provide a way of showing the stars. In the tower, a solar telescope took the place of the dismantled refractor. It was used for training students, and actually uses the lens from the refractor. (Nowadays, when astronomers at the observatory have discovered that history can be put to use, plans for a reassembly of the telescope have been discussed in connection with a future relocation to new observatory buildings. In such a new situation, the Lund astronomers seem to feel that a vintage instrument would be a valuable exhibition piece.) What was once, in Dunér's time, a research grade instrument – especially after Dunér fitted the telescope with spectroscopes – has passed, via a use for popular science, into a dismantled state, some of the parts being used in a training instrument. The instrument has been adapted to various contexts of astronomy.

[341] Knut Lundmark, *Världsrymdens liv* (Uppsala, 1926).

which was an enlargement of the previous book.[342] Lundmark was not really satisfied with it, however; the finished product was a bit too unsystematic in its mixture of literary imagination and astrophysical views, he thought. He also wished he had been more critical of Arrhenius and noted that the book had a slightly too Flammarionian flavour.[343] A projected second part never was published, after troubles with the publisher.

In the early 1930's, Emanuel Linderholm started Sveriges Religiösa Reformförbund (the Swedish Association for Religious Reform). It published the journal *Religion och kultur* (Religion and Culture). Linderholm argued that theology needed to be updated on modern developments in various fields, science being one of them. Lundmark published an article on the large-scale structure of the universe, where he supported Charlier's views of an infinite universe.[344] Besides publishing other articles in *Religion och kultur*, he was also approached by a committee organised by various groups connected to the Swedish church (Svenska kyrkans diakonistyrelse, Sigtunastiftelsen and others). Its aim was to make religious issues more visible in the daily debate. Lundmark's role was to publish a text on the subject of religion and *Weltanschauung*.[345]

Lundmark's various writings in popular science – articles, books, later on also in radio form – made him an almost famous figure in Sweden. That activity took quite a lot of time from his other activities and duties as a research astronomer and head of the department of astronomy. There were several reasons for this. One motivation was a sense of need for science in culture. Astronomy and other sciences should be visible in the culture at large, Lundmark in effect argued, by committing so much time to those activities. Another reason was financial. Like many other students of the time, Lundmark acquired debts during the years at the university to banks and also to fellow students (via the Swedish system of loans where students guaranteed each others' loans in the banks, *växlar*). A web of economical transactions tied him to other scientists of his gen-

---

342 Knut Lundmark, *Livets välde: Till frågan om världarnas beboelighet. I.* (Stockholm, 1935).

343 Knut Lundmark to Per Collinder September 21 1934, Lundmark's correspondence, LUB.

344 Knut Lundmark, "Ignorabimus? Till frågan om världens oändlighet", *Religion och kultur* vol 1 (1930), 89-100.

345 Manfred Björkquist and Gösta Lindeskog to Knut Lundmark n.d. November 1933; Gösta Lindeskog to Knut Lundmark 5 November 1934, 4 September 1935, 17 January 1937, 25 September 1938, 5 October 1938, Lundmark's correspondence, LUB. Lundmark never published the text.

eration.[346] Financial problems did not necessarily disappear after he had become professor. Loans taken by fellow students and guaranteed was a continuous threat. Such a thing happened in 1934 when such a friend from student days died, leaving among other things a loan in a bank with Lundmark as guarantor. Lundmark needed money fast. The book on life in space, *Livets välde*, helped increase the finances somewhat (and also seems to have been published in part because of that), and he pushed his editors to pay him quicker for the stream of articles that flowed from his typewriter. Lundmark published articles on just about anything in his wide sphere of interests, from Strindberg to stars, from botany to astrobiology.[347]

While he early on had been working with the Swedish Astronomical Society that catered to a more advanced type of astronomy enthusiasts, Lundmark felt more ill at ease with that society. Not the least since he had come to Lund in the southern part of Sweden, he participated less frequently in the activities in the national society, which to a large extent had a focus at Stockholm/Uppsala. Lundmark gradually became distanced from the Swedish Astronomy Society. Instead, he founded a local astronomic society, the Tycho Brahe astronomy society, in 1937. It was tied to the Lund observatory, but had its meetings in nearby Malmö. The Tycho Brahe society published the yearbook *Cassiopeia* from 1939 and had a high level of activity.

The 1930's saw an increase in Lundmark's activities outside of academic astronomy. Slowly Lundmark migrated from being a research scientist and institution leader in the first place to being more and more a writer and populariser with many connections to the culture at large. During the 1940's this process appears to have continued. It was during that decade Lundmark published his magnum opus (at least as popularisations go) *Nya himlar*, a monograph on Strindberg. Lundmark's career more and more consisted of activities outside of astronomy, a science he once had claimed he did not want to leave because he loved it.[348]

---

[346] In the early 1920's he had debts of over 16 000 kr. Sten Asklöf to Knut Lundmark 10 January 1923, Lundmark collection, LUB.

[347] Knut Lundmark to Ingeborg Lindgren 1 November 1934, Knut Lundmark to Helge Lindberg 12 November 1934, Lundmark's correspondence, LUB. Bertil Lindblad had similar troubles. He was guarantor to someone who went bankrupt and rapidly Lindblad needed money; Lundmark helped him. Bertil Lindblad to Knut Lundmark 21 August 1928, Lundmark's correspondence, LUB.

[348] Knut Lundmark to Östen Bergstrand 16 December 1921, Bergstrand's collection NC736, UUB.

Lundmark on the Periphery

Lundmark was marginalised in astronomy during the 1930's and 1940's. His large project of cataloguing nebulæ soon got into trouble. It did not make the Lund observatory into the obligatory passage point in international nebular astronomy that Lundmark had envisioned. Gradually he became more peripheral in Swedish as well as international astronomy. That is one factor that can explain his activities in popular astronomy. Slowly he moved from a rather central position in the 1920's (worked internationally, became professor) to a more peripheral position with a weaker network.

Lundmark felt left out from the central parts of Swedish astronomy, a feeling that was grounded to some extent in a self-esteem that was sometimes rather low, but also in a difference in scientific style. Several astronomers thought that Lundmark, with his stronger interest in broad issues and popularisation, lacked strictness. They belonged to different thought collectives.

During the 1930's he tried to start an international journal devoted to nebular astronomy. He saw it as "an archive, mainly for longer empirical papers and theoretical or analytical works", with surveys of the literature etc. He had a voice of support from Harlow Shapley.[349] Discussions had been started with the publisher Almqvist & Wicksell on the possibility of launching the journal. *Acta Metagalactica* he called it. Towards the end of the thirties Lundmark hoped that the project could start. In the first issue, Lundmark thought about publishing the Messier catalogue of nebulæ together with a survey of nebular observations from old times.[350] It never materialised.

His position on the periphery became obvious when the International Astronomical Union held its general assembly at Stockholm in 1938. Lundmark spent the early part of that summer in the US, among other things connected with the celebrations of the Swedish colony founded in Delaware 300 years earlier.[351] This is one reason why he was rather disconnected with the work in organising the general assembly – another reason was his more general placement on the periphery of Swedish as-

---

[349] Knut Lundmark to Bertil Lindblad 18 September 1933, Lundmark's correspondence LUB.

[350] Knut Lundmark to Per Collinder 19 July Lundmark's correspondence, LUB.

[351] Per Collinder to Knut Lundmark 25 March 1938, Lundmark's correspondence, LUB.

tronomy. Bertil Lindblad and the Stockholm/Saltsjöbaden observatory was, together with the Uppsala observatory, more central in Swedish astronomy.

Lundmark tried to organise a trip to Ven and the Tycho Brahe remnants for the participants of the congress. The idea was that when so many prominent international astronomers were visiting Sweden, it would be a good idea to have them come and visit one of the few really significant places in the history of astronomy that existed in Sweden. In mid-July of 1938 Lundmark asked Lindblad to send him a list of all the astronomers that were participating in the conference. Earlier he had got an indication from Lindblad that there was no possibility of obtaining funds from the conference budget for an official trip, so Lundmark decided to go on with his own plans.[352] Ven being close to Lund, it seemed an easy task for Lundmark to make the group visit the ruins, and it would also put Lund on the map for the international astronomers.

Lindblad was quite critical of the idea. Lundmark was too late: he should have described his plans earlier and Lindblad also accused him of being difficult to get in touch with during the months he had spent in the US. Lindblad also proposed a co-operation between Lundmark and Elis Strömgren, professor of astronomy at Copenhagen; Strömgren had earlier voiced an interest in doing something on Ven during the general assembly.[353]

Lundmark became angry. He wrote that "you up there [to the north] had really showed no real interest in making Lund part of this year's [general assembly]". He also voiced his opinion of how he felt as not being quite part of the Swedish astronomical community:

> I thought that the Swedes present during [IAU:s general assembly in 1935] in Paris with one exception behaved very distantly and strange, and that it was clearly shown that I did not belong to your circles [...] This is a minor matter, but I say this to indicate that I cannot force myself upon you. Therefore, I have chosen to not have anything to do with the organisation of this congress.

Lundmark felt like an outsider in the astronomical establishment in Sweden. He had not been present during the planning of the congress. He had thought of making a smaller, private excursion to Ven and now Lindblad had the guts to propose how Lundmark should be handling the details of the excursion. "But how, how???" Lundmark burst out. Lind-

[352] Knut Lundmark to Bertil Lindblad 18 July 1938, Lindblad's papers, F2, KVA.
[353] Bertil Lindblad to Knut Lundmark 20 July 1938, Lindblad's papers, F2, KVA.

blad proposed detailed changes to Lundmark's plans (including, for example, a whole series of excursions to Ven under the guidance of assistant astronomers from Lund) that were too costly, they came too late to plan and the assistant astronomers of Lundmark's observatory could not be expected to devote considerable amounts of time to the venture. "Your letter has in several aspects touched me in a most unpleasant manner" Lundmark wrote. He had not at all been difficult to get in touch with during his visits to the US, he answered. In a display of kindness he offered to "avoid ruining your congress by exhibiting this unpleasant story during the congress".[354]

Lindblad answered that he had got the impression from Lundmark's first letter that the planned trip to Ven was a project on quite a large scale, and hence it was a bit late to start talking about it in late July. No proposal for a trip to Ven during the IAU conference had reached the conference headquarters during the three years the conference had been in the planning stage. Lundmark had been treated just like any other member of the National Committee for astronomy "and I can guarantee that every proposal from you would have been dealt with in a most serious manner." There were no funds left to do a trip with the conference to Ven, unless the planned trips to Uppsala or Saltsjöbaden were cancelled – Lindblad clearly demonstrated that a visit to Ven (and with it, Lund) came number three, hence his view of the pecking order of astronomy in Sweden. Lindblad was also hurt by the exchange of letters: Lundmark's "words about loyalty and sacrifice of time hurt me somewhat", since the staff at Saltsjöbaden had had to sacrifice lots of time for the conference. [355] In his following letter, Lundmark kept talking about his peripheral place in astronomy. "No signs were shown at [the previous IAU general assembly] in Paris that there would be any co-operation between Lund and Stockholm in this matter. It was as great a surprise for me as for the general public, when an invitation to come to Stockholm was issued."[356] The solution to the problem was that Lundmark cancelled the plans for a trip to Ven.[357]

[354] Knut Lundmark to Bertil Lindblad 21 July 1938, Lindblad's papers, F2, KVA. The proposed cooperation with Strömgren in the visit to Ven was especially inconvenient for Lundmark; by 1938, he had come to regard Strömgren as a tyrant, and refused to let Lindblad and Strömgren organise the event. Knut Lundmark to Per Collinder July 28 1938, Lundmark's correspondence, LUB.

[355] Bertil Lindblad to Knut Lundmark 23 July 1938, Lindblad's papers, F2, KVA.

[356] Knut Lundmark to Bertil Lindblad 24 July 1938, Lindblad's papers, F2, KVA.

[357] Knut Lundmark to Bertil Lindblad 28 July 1938, Lindblad's papers, F2, KVA.

He changed his mind and was now determined to talk about "this horrible business" with visitors to the general assembly. "Of course I will explain things to the Americans."[358] He thought about distributing a note in English to the American astronomers and to some English and German delegates.[359] After the IAU general assembly, he wrote to Ferdinand Neubauer at the Lick observatory, saying that Sweden was too small a country and that he longed for California. He also claimed to be glad that neither Bergstrand, Lindblad nor von Zeipel was elected the IAU:s president.[360]

Just like the case with the conference in 1904, when the Astronomische Gesellschaft met at Lund, a visit to Ven was an event that the Swedish and Danish astronomers thought would be of interest to the foreign guests. Astronomers from both countries laid claim in 1938 on the island and wanted to show it to the international astronomical world. The incident with the Ven trip also clearly shows how marginal Lundmark's position had become in Swedish astronomy.

When Östen Bergstrand was to be celebrated with a *Festschrift* it was thought, by some Stockholm astronomers, important that Lundmark did not lead the project: to prevent him from producing a book that was inaccurate, too wide-ranging in topics and dealt with subjects like life in the universe.[361]

Lundmark was never elected to the Royal Academy of Sciences at Stockholm. Instead, he saw Walter Gyllenberg and Gunnar Malmquist being elected in 1939; an astronomer that he had competed successfully against for the Lund professorship was elected, not him. Lundmark took it rather hard.[362] (After the time period covered here, in 1949, another election to the Academy went to Carl Schalén. And in connection with this later election, we find some mentions by Lindblad on the qualities of Lundmark. Lindblad states that Lundmark had not produced so much in the way of science and that his work in the field of popular astronomy was of the wrong "mentality", and the content and literary style was "partly [...] quite silly".[363])

[358] Knut Lundmark to Per Collinder 28 July 1938, Lundmark's papers, LUB.

[359] Knut Lundmark to Erik Holmberg 28 July 1938, Lundmark's papers, LUB.

[360] Knut Lundmark to Ferdinand Neubauer, 19 August 1938, Lundmark's correspondence, LUB.

[361] Sten Asklöf to Carl Schalén, 14 January, 1936, Carl Schalén's papers, LUB.

[362] Anita Sundman, *Den befriade himlen*, 94.

[363] Copy of a letter from anon. (but it is clear from the content that it is written by Lindblad) to unknown 26 January 1949, Bertil Lindblad's papers, KVA, B1:1.

An instructive example of the difference between Lundmark, Lindblad, and Östen Bergstrand is evident in a book on modern science published in 1937 by the newspaper *Svenska Dagbladet*.[364] The work was a summary of the state of science, with contributions from leading Swedish scientists.

In his article, Lundmark ventured to discuss questions in a typical manner.[365] He stated that a formal distinction between living entities and dead matter would be impossible to uphold; stars can be "said to be something akin to [...] living creatures that reach certain basic definitions of life processes."[366] Stars do not have minds, but they can, like lower forms of life, reproduce, absorb nutrition and regulate their growth. The thought was not new, for it had once been put forward by Aristotle. "Once again modern science might have to return to the philosopher from Stagira, which is not something to be ashamed of."[367] Besides these thoughts, Lundmark reflected on science's view on the big questions. Science had become more humble than during "more optimistic times coloured by enlightenment"; new domains had been opened up, which showed that science hardly could be said to be close to the ultimate answers; the scientific advances had led to a feeling of restraint among scientists. For Lundmark, the result was a religious feeling of respect for the cosmos, he almost saw a creator behind the plan of the world.[368] In this text Lundmark also lamented what would later be called the problem of the two cultures.

Lindblad's text was different; he did not draw any large metaphysical conclusions nor make any pronouncements on the sad lack of humanistic *bildning* among his fellow scientists. Lindblad saw a break in the history of astronomy around 1900, when classical astronomy gave way to newer methods of investigating the cosmos. He explained that an important task was "to gain knowledge of the state of matter in the parts of a star that emit radiation to us, through the empirical results of spectral photometry".[369] Other important issues faced by astronomers included the

---

[364] *Olösta gåtor och aktuella problem: sjuttiosex uttalanden i Svenska Dagbladets enquête om naturforskningens nuvarande uppgifter och mål* (Stockholm, 1937).

[365] Knut Lundmark, "Enhet och mångfald i universum", *Olösta gåtor och aktuella problem*, 165-175.

[366] Ibid., 166.

[367] Ibid., 167.

[368] Ibid., 168f.

[369] Bertil Lindblad, "Varifrån och varthän går utvecklingen?", *Olösta gåtor och aktuella problem*, 157-160.

stellar energy problem, studies of the Milky Way and the extragalactic stellar systems, as well as modern cosmological results showing the universe to be in rapid development, rather than stable. The only part of his text where Lindblad left a style that was a sombre catalogue of the advances of modern stellar astrophysics was when he discussed the issue of why one would pursue science at all. Though science had been instrumental in averting material barbarism, material advances ought not to be the main reason for pursuing science. He complained that many people gave utilitarian reasons for science instead of the great role played by the sciences in "the sanitation of human culture through their immutable demands for truth."[370]

Östen Bergstran's text exclusively dealt with a description of the main advances of stellar astronomy.[371] It is interesting to note how Bergstrand distributed his praise for astronomers in the younger generations. Two fields were of particular importance. One was Lindblad's and Oort's discovery of the rotation of the Milky Way, "one of, if not the most, important advances in astronomy since the turn of the century." The other was the field of interstellar absorption, where "C. Schalén of Uppsala is in the front line". No other astronomers in this field were mentioned. Studies of extragalactic objects were still very uncertain, "they belong almost completely to the research of a future era".[372]

*

The work on the Lund General Catalogue of Nebulæ soon ground to a halt. In several of the yearly activity reports from Lund in the latter part of the 1930's, it is not mentioned at all. Lundmark did other things: popular astronomy, studies on novæ and supernovæ, history of astronomy, journalistic work, studies on Strindberg, botany. In the yearly report of 1942-43 the project is however mentioned again. Lundmark tells that during the year he has collected material to an introductory volume that would provide a history of nebular astronomy from the oldest times to the Herschels.[373] Such a study was never published in a scholarly form. His *magnum opus* of popular astronomy, *Nya himlar: Från stjärnkunskap-*

[370] Ibid., 160.

[371] Östen Bergstrand, "Nya vidder för celest forskning", *Olösta gåtor och aktuella problem*, 27-30.

[372] Ibid.

[373] Yearly report of the Lund observatory, *LUÅ* 1942-1943, 41-44, 42.

ens gryning till vintergatornas vintergata published in 1943, was such an historical introduction. But it was written for the layperson and the educated public, not for the astronomer who needed a historical treatise on nebular astronomy.

Lundmark never got the amount of resources that are necessary to run such a project. He could use the computers that had been at the observatory since the days of Charlier. He could steer some of his graduate students toward nebular astronomy, but the project never got large external funding.

When a copy of Charlier's *A Plan for an Institute for Theoretical Astronomy* was bound by Lundmark, he added a page where he gave a hint of what he felt about his scientific achievements:

> This work is written by C.V.L. Charlier. When it was published, I am afraid it did not get the recognition that I undoubtedly think it deserves. This copy does not look all too nice, but it came as a gift from Sven Wicksell's library via Carl-Erik Quensel, who had bought Wicksell's library. I thought when it came here that we might as well bind it as I could not find a copy in the observatory library at the moment. In its saddening state it is a memory, both of Wicksell and Charlier; it also reminds us, by the very state it is in, of the limitations of life and human spirit. It shows that it is not enough to have a grand plan. If the right circumstances are not present, well, then the genius and originator of the plan can say goodbye at once. Much of the advances of science is because of pure chance.[374]

In 1942 he was bitter of the changing attitudes toward nebular classification and the resources that he had not been able to tie to the project. The *Lund General Catalogue* had got no sustained support, neither from Swedish nor international astronomy. Astronomers had not supported the project and they had not been interested in sending their data to Lundmark's proposed clearinghouse for nebular astronomy in Lund. Lundmark never got the satisfaction of seeing large parts of the world's empirical material on nebulæ pass through and being treated at his institute. He also was unsuccessful in getting the support of research establishments like Lund University or the Royal Academy of Sciences. He tried to make use of his contacts with international astronomy, such as sending pupils to observatories where telescopes, photographic plates and fair weather were present.

---

[374] Text dated July 7 1942, bound together with a copy of C.V.L. Charlier, *A Plan for an Institute for Theoretical Astronomical Research* (Lund, 1913), copy in Lund observatory's library.

Instead of getting large amounts of pure data passing through the Lund observatory's project, Lundmark instead had to deal with published material. He had to massage the data. To make the data useful, it had to become combinable and comparable. Lundmark's task was to produce data that could be used by other scientists.

Lundmark's project was a pragmatic reaction to the fact that he did not have access to large telescopes at the Lund observatory. By pursuing the *LGC* he also used the resources present at the Lund observatory, resources that his predecessor as professor had built up. Charlier's work in stellar statistics was a local scientific culture that Lundmark could continue to use in his catalogue project. When he arrived at Lund, five computers were employed there, which was a resource that was missing at other observatories in Sweden; Hugo von Zeipel looked upon the Lund resources in computing as something that was desirable for the other observatories in Sweden.[375] Lundmark could hook up his nebular catalogue and classification project to this local resource that was somewhat stable – it could not easily be exchanged for big telescopes, for example. His project of nebular classification and cataloguing was made to fit with the existing local culture of astronomy.

Lundmark's project was also an outcome of his style. He claimed that it was not important to deal purely with mathematics to do real work. He was more interested in gaining knowledge in various parts of astronomy. He had read widely and perhaps knew the astronomical literature better than most astronomers, a knowledge that fitted well with his catalogue project. In letters to his former teacher and thesis supervisor Bergstrand, he lamented that his way of working did not fit in with what many perceived as cutting edge and with the concept of the genius. Thus, early in his career he was aware of his personality. He also felt that Bergstrand favoured Lindblad over himself, citing friends who had heard Bergstrand claim that Lindblad's work was revolutionary. In fact, Lundmark perceived Lindblad, rather than himself, as Bergstrand's pupil and successor. He thought that Bergstrand felt that Lundmark's work was too hypothetical instead of empirical, and identified Lindblad's style as closer to the cautious and empirical ideal that was Bergstrand's. He also identified a difference in personality as well as in scientific matters.

[375] Yearly report of the Lund observatory, *LUÅ* 1928-1929, 35-38, 38; Hugo von Zeipel, "Några önskemål för svensk astronomisk forskning", *PAT* 1922, 62-64, 63.

Lundmark said that he had been hindered by low self-esteem, and a lack of concentration of work.[376]

Knut Lundmark became even more peripheral in astronomy, and this increasing peripherality is one explanation for the increasing activity he developed outside of academic astronomy. The extramural activities were rather intense during the years studied here, and if anything increased after 1940. Knut Lundmark was during the 1940's and 1950's known not as the leader of a major international centre of calculation and astronomical datahandling, a leading journal in extragalactic astronomy, a big and active observatory in Lund, nor as a member of the Royal Academy of Sciences, but as a writer on subjects like astronomy in the Bible, August Strindberg, the history of astronomy, and the history of Piteå in northern Sweden.

[376] Knut Lundmark to Östen Bergstrand 16 December 1921, 18 September 1924, Bergstrand's collection NC736, UUB.

FROM UPPSALA AND STOCKHOLM TO THE STARS

DURING the 1920's and 1930's, a group that had developed around Östen Bergstrand gained ground and soon became the dominant force in Swedish astronomy. Östen Bergstrand's pupil Bertil Lindblad eventually became the centre of this group. The group specialised in a kind of astronomy that did observational astrophysical studies of stars and the stellar system. Its scientific aims were centred on measuring properties of stars that would elucidate the structure of the Milky Way system of stars. The group developed methods that borrowed from advances in modern astrophysical practice and theory.

Central to the group was the build-up of a new institutional home, the Stockholm observatory at Saltsjöbaden. It became possible after a large donation from the Knut and Alice Wallenberg foundation, an important supporter of culture and science in Sweden. The observatory, under the leadership of Bertil Lindblad, worked on large programmes of observational astrophysics. Instruments were chosen that largely were of the modern astrophysical type for stellar and nebular studies.

Modernisation of Stellar Astronomy: Lindblad and his Group

Charlier's programme in stellar statistics – it was often called the Lund school – began loosing ground during the 1920's. Internationally, astronomy went in other directions, both practically and theoretically. The cosmology presented by the stellar statisticians was being criticised; several astronomers began discussing the stellar system in sizes up to 100 000 to 300 000 light years, not 20 000.

At stake was not only the results of astronomical research, but also the ways to do science, astronomy's practice and material culture. A more modern form of stellar astronomy began to gain ground that was theoretically informed, without dealing exclusively with theory. The theoretically-informed observational astronomy was a style on the march forward; compared to the styles of empirical collection (Lundmark) and statistical treatment (Charlier), it became the thought style of a new school of astronomy. Non-observational programmes, like the statistical works of Charlier or the celestial mechanics lost ground, together with Lundmark and his way of doing astronomy, which was a bit like natural history. How this style of astronomy came to get such a prominent place in Swedish astronomy will be discussed in this chapter. One of the main characters was Bertil Lindblad.

New physics and astronomy were being integrated into a program in stellar astronomy. Lindblad and his colleagues worked on various ways of getting to the structure of the stellar system, the properties of the stars, how stars and dust were being distributed in the stellar system. The areas of research were thus in some aspects similar to the Lund school. Just like the Lund school, the new way of doing stellar astronomy was concerned with charting the structure of the stellar system by using empirical data. But there were differences. The Lund school had been doing astronomy based on statistical analysis of properties of the stars: positions, brightness, velocities, the apparent density of stars on the sky, but they had in a quite small degree been doing astrophysical work. Charlier and the other stellar statisticians were more geared toward mathematics proper than physics. Lindblad and his group worked with getting methods for determining the properties of stars by using the more modern astrophysical methods. The difference in scientific style is a mirroring of a changing astronomy between 1905 (when Charlier started doing stellar astronomy) and about 1920-1925, when the Lindblad group formed. The stellar astronomy field was strengthened by input from other areas. Stellar astronomy changed style; it was perceived as a modernisation.

One difference between the Lund group and Lindblad's group concerns the role of the observer. Charlier and his pupils had worked with analysis of observational material from other observatories. Catalogues of stellar data were used in the analysis, in some cases photographs of the sky. The Lindblad group was more into doing the observations themselves. In the Lindblad group there was an idea that astronomical results should be connected with observations made by the astronomer himself.

In contrast, Charlier's school worked on developing the mathematical treatment of rather basic observations: brightness, position, radial velocity, proper motion. Working out new observational methods or developing existing methods were deemed as good achievements in the Lindblad group; doing large-scale observational programs was looked upon positively.

An important person behind the growth of the Lindblad group was Östen Bergstrand. Bergstrand's empirical mode of doing astrophysics was a local style of science that was being transmitted to Bergstrand's pupil Lindblad.[377] Bergstrand in turn had Nils Dunér as a teacher, who also was an observational astrophysicist. The empirical style of the Lindblad group was connected with the working out of observational methods that were suitable for the situation as regards instruments at Uppsala and later Stockholm/Saltsjöbaden. The Saltsjöbaden and Uppsala stellar astronomers did not see themselves as simple fact-collectors, for them the theory behind the instruments and methods were important.

Research on the Milky Way was done in roughly three phases. In phase 1, the sky is observed with a telescope (that is sometimes equipped with auxiliary instrumentation like spectrographs, filters &c.), resulting in photographic plates of stars or stellar spectra. Phase 2 involved the reduction of the photographic material and the extraction of the data on the stars from the plates. In phase 3, an astronomer uses data from phase 2 for getting information on the positions of the stars in space, their composition, the dynamics of stellar systems &c. The data used here can be produced by the astronomer himself in phase 2, or it could be produced somewhere else. But the very fact that data could be transported for analysis at some other geographical location and in another institutional setting did not of course imply that it would be transported. The observatory that had produced the observational material was not always interested in sending away the photographic plates: the staff at the observatory should analyse it themselves, or at least do a reduction and publish it as a catalogue. Phase 1 was relatively simple to perform for the observatory that had the necessary prerequisites for such work, but it gave quite little status in the astronomical world. Reducing raw data to published tabular data was a merit, as was analysing such data.

---

[377] In an introduction to the Lindblad group at Uppsala, Lindblad explicitly traced their ancestry to Bergstrand's development of the effective wavelength method. Bertil Lindblad, "Spektralfotometriska undersökningar vid Upsala observatorium", *NAT* ns vol 7 (1926), 41-50, 41.

The astronomer or the head of the observatory had to make a choice of whether or not to observe the sky with equipment at the home observatory. In contrast to Charlier and Lundmark, Lindblad chose the first alternative. They worked at getting a self-supporting institute, where photographic data was to be produced, reduced and analysed at the home institute, in some cases through the use of instruments and methods of observation and analysis that had been developed there as well. This meant that Lindblad and his group did not have the problems connected with convincing other observatories to part with their observational data, like Charlier and Lundmark had. On the other hand, this choice entailed problems connected with developing and running an observatory with adequate instrumentation, a staff that was skilful in developing observational methods, and battling the Swedish weather. (It is not altogether certain that Lindblad made a choice. As will be argued below, Lindblad's career and the fate of his group were shaped both by the agency of the astronomers and the surrounding institutional, economical, and political circumstances.)

Lindblad and his group worked in a field where methods for measuring the luminosities and distances of stars were very important. These fundamental areas were worked upon with methods and instruments taken from the arsenal of methods of modern astrophysics. Lindblad defined modern astrophysics as spectral analysis and stellar photometry.[378]

If a star's luminosity is known, the apparent brightness of the star can be converted to the distance of the star. From about 1910 astronomers started discussing the notion of stars having different luminosities even if they were of the same spectral class. A search for luminosity criteria ensued. If stars of the same spectral class could have different luminosities, there might be some details in the spectra of the stars that would enable astronomers to distinguish between a dwarf and a giant star. The spectroscopists Walter S. Adams and Arnold Kohlschütter showed in 1914 that there were vague differences in the spectra of red stars that enabled them to distinguish between giants and dwarfs.[379] It became possible to measure

---

[378] Bertil Lindblad, "Våra dagars astronomiska huvudproblem. Intryck från den Internationella Astronomiska Unionens kongress i Cambridge, U.S.A. i september 1932", *PAT* vol 13 (1932), 163-178, 173.

[379] Walter S. Adams and Arnold Kohlschütter, "Some Spectral Criteria for the Determination of Absolute Stellar Magnitudes", *ApJ* vol 40 (1914), 385-398. Cf. Hearnshaw, *Analysis of Starlight*, 216ff.

distances to stars from spectra. More astrophysicists started working on spectroscopic ways of determining stellar luminosities and distances.

In the following, we will see how the Lindblad group was developed and how it originated at the Uppsala observatory, and we will see how the group manoeuvred to get instruments that were of use in the group's work. We will discuss the work the Milky Way group connected with Lindblad did in astronomy and how it developed into a leading position in Swedish astronomy as well as gaining some success on an international scale.

*

Bertil Lindblad (1895-1965) came to Uppsala as a student in the autumn of 1914.[380] His studies in astronomy were under the professors Östen Bergstrand and Hugo von Zeipel.

Lindblad was from early on interested in the measurement of stellar colours. His interest was an effect of the local practice at the Uppsala observatory. Östen Bergstrand had started out in photographic classical astronomy (stellar parallaxes) but in about 1907, Bergstrand begun to measure stellar colours by photography. He had begun using methods that allowed stellar colours to be measured with the comparatively small instruments available to the Uppsala observatory's budget. He measured the colours of stars by taking plates with a coarse grating in front of the telescope. The so-called effective wavelength (a measure of stellar colours) was made on the plate.[381] Bergstrand had worked on getting the method of effective wavelength determination adapted to a wide-field telescope. The point of such an instrument was that it could observe many stars at the same time. Bergstrand hoped that the method of effective wavelength would allow the small Uppsala observatory to compete with larger observatories, who ordinarily observed spectra with spectrographs; Bergstrand hoped to be able to produce similar data with a much smaller instrument (see chapter two above).

Lindblad soon started to work along the lines introduced by Bergstrand. He became interested in trying to find luminosity criteria for stars by using Bergstrand's technique. Lindblad tried to find luminosity criteria that was coupled to the distance between what he called the minimal

---

[380] Biographical data on Lindblad is in Yngve Öhman, "Lindblad, Bertil", *SBL*; Jan H. Oort, "Obituary Notices: Bertil Lindblad", *QJRAS* vol 7 (1966), 329-341.

[381] Knut Lundmark, "Carl Östen Emanuel Bergstrand", *SBL*; chapter two, above.

wavelength and the effective wavelength, which would enable him to determine stellar absolute magnitudes, and hence, distances to the stars.[382] Some time later, when his method was published, he fought a lesser dispute regarding the qualities of the method with Knut Lundmark and W.J. Luyten.[383] Lundmark and Luyten had argued against Lindblad's method. It might work in theory but not in practice. Strong feelings were sometimes voiced in letters from Luyten to Lundmark; Lindblad was regarded by Luyten as not having a very nice tone in the debate.[384]

After Lindsblad's PhD thesis was defended, Lindblad came to Mount Wilson, one of the most important observatories of the new era. He was not given direct and frequent access to the large telescopes but he had some use of the 60 inch reflector. Instead, he used a smaller photographic telescope that had an objective prism. With this telescope, it was possible to photograph the spectra of many stars at the same time.[385]

During the time in California, Lindblad developed a method of determining stellar luminosities from spectra. He showed that two features in stellar spectra changed with changing luminosity: the width of the Balmer lines, and the strength of the CN band. Stars of high luminosity had narrow Balmer lines, faint stars like white dwarfs had very broad lines. The CN band visible in red (cool) stars varied with the luminosity. By studying these spectral details it was possible roughly to determine the absolute magnitude for the star, which then could be used to determine the distance to the star. Since spectra of stars are easily observed compared to the measurement of parallaxes, such a method of obtaining spectral parallaxes, as it was often called, meant that astronomers could ex-

---

382 Bertil Lindblad, "On the Use of Grating Spectra for Determining Spectral Type and Absolute Magnitude of the Stars" *ApJ* vol 49 (1919), 289-302.

383 Knut Lundmark and W.J. Luyten, "On the Determination of the Colour-Equivalent of a Star, with Special Reference to the Effective Wave-Length and its Relation to Spectral Class: A Review of the Different Methods", *MNRAS* vol 82 (1922), 495-509; Bertil Lindblad, "Note to Dr. Knut Lundmark and Dr. W.J. Luyten's Paper on the Determination of the Colour-Equivalent of a Star", *MNRAS* vol 83 (1922-1923), 97-98; Knut Lundmark and W.J. Luyten, "Note on the Determination of Absolute Magnitude from λ_e and λ_m", *MNRAS* vol 83 (1923), 470-474; Bertil Lindblad, "On the Intensity-Distribution in Short Grating Spectra and Objective-Prism Spectra as a Function of Spectral Type and Absolute Magnitude", *MNRAS* vol 83 (1923), 503-510.

384 W.J. Luyten to Knut Lundmark February 24, February 25, March 16, October 6 1923, Lundmark's correspondence, LUB.

385 Bertil Lindblad to Knut Lundmark October 27 1920, Lundmark's correspondence, LUB.

pand their knowledge of stellar distances. The classical method of trigonometrical parallaxes only reached out a very short distance in space, since the parallactic angles for all but the very nearest stars are very small.[386]

Writing home to his mentor Östen Bergstrand from Mount Wilson in the summer of 1921, Lindblad said that he had reached an end in his search for luminosity criteria in spectra with his discovery of the criteria he had developed at Mount Wilson. The next step would be to start using the criteria on larger numbers of stars when he came home, and preferrably together with some other astronomers. He had in mind a group at Uppsala working on applying and developing his method further. A point with the CN method was that it worked well with smaller instruments, Lindblad wrote Bergstrand. "The very faintest stars we might have to leave to the reflectors, but down to the 8th and 9th magnitudes there is much work to do!" Lindblad had found a method of measuring the luminosities of stars that did not hinge on him having access to very large telescopes, and thus would make it possible for smaller-scale instruments at a typical european university observatory, like Uppsala, to continue to produce measurements of the structure of the Milky Way system. The method carved out a niche for smaller instruments. It could be used for measuring stellar luminosities with survey-type instruments with large fields of view, hence getting results for many stars at once. The method was suitable for mass-production methods and also suitable for observatories with lesser resources and instruments. In this regard, his criteria was good news for astronomy in Uppsala and Sweden.[387]

When Lindblad returned to Sweden he formed a small school of astronomers that worked in stellar astronomy, with the young astronomers Carl Schalén and Yngve Öhman as core members. This group of astronomers were aiming at getting to the distribution of stars in space, the same general goal as the stellar statistical school. But where Charlier and his colleagues tried to get at the distribution of stars in space by analysing stellar data statistically, Lindblad and his group tried to develop methods for discerning the distances to individual stars. They worked on

---

[386] Hearnshaw, *Analysis*, 270f; Bertil Lindblad, "Spectrophotometric Methods for Determining Stellar Luminosity", *ApJ* vol 55 (1922), 85-118; Bertil Lindblad, "De spektroskopiska metoderna för bestämning av stjärnornas avstånd", *PAT* vol 3 (1922), 33-42.

[387] Bertil Lindblad to Östen Bergstrand, 29 July 1921, Bergstrand's papers, NC736, UUB.

developing luminosity criteria that would enable them to determine the distances to a star by analysing the spectrum of star. In a series of papers, Lindblad, Schalén and Öhman published results obtained with the method and developments of the method.[388]

One feature of the methods Lindblad and his group had developed was that it worked with small-scale spectra of stars. Thus, it was not necessary to have access to the large instruments that could produce high-dispersion spectra of stars: the Lindblad group could instead work with domestic instruments; it used the Uppsala astrograph with an objective prism. The small spectral resolution of the Uppsala instrument was offset by the nature of the Lindblad method. The method was further developed: instead of estimating the intensity on the spectra, the spectra on the photographic plates were measured with a photometer. Using a photometer to measure the plates made the method more accurate. The group of Milky Way astronomers developed the method towards a more quantitative style. A machine measured the spectra and plotted the spectra automatically. In the place of an individual astronomer's ability to recognise patterns there was a machine working on the spectra.[389]

While Schalén, Öhman and others worked empirically, Lindblad himself also took up more theoretical studies of the stellar system. He had got a group going on observational projects; now he himself observed less and worked more with theoretical aspects of stellar astronomy. From around 1925 Lindblad studied the dynamics of our stellar system as well as other systems. His theoretical studies was directed towards questions about the motions in the stellar system.[390] Eventually this led to his discovery of the rotation of the stellar system. He also explained Kapteyn's discovery of stellar streams as an effect of the stars' orbits in the stellar system.[391]

---

[388] For a bibliography that gives the main publications by the Lindblad group from 1922 to 1932, see Bertil Lindblad & Erik Stenquist, *On the Spectrophotometric Criteria of Stellar Luminosity* (*Astronomiska iakttagelser och undersökningar å Stockholms observatorium* vol 11 no 12) (Stockholm & Uppsala, 1934), 3, note 1.

[389] Bertil Lindblad, "Spektralfotometriska undersökningar vid Upsala observatorium", *NAT* ny följd vol 7 (1926), 41-50.

[390] Bertil Lindblad, "On the Cause of Star-Streaming", *ApJ* vol 62 (1925), 191-197.

[391] Cf. R.W. Smith, "Studies of the Milky Way 1850-1930: Some Highlights", in Hugo van Woerden, Ronald J. Allen, W. Butler Burton eds., *The Milky Way* (IAU Symposium no 106) (Dordrecht, 1985), 43-48, 51ff; Dimitri Mihalas, *Galactic Astronomy* (San Francisco, 1968), 119f; Bart J. Bok & Priscilla F. Bok, *The Milky Way* (Philadephia 1941; 1945), 92-104.

Lindblad also took up the question of the motions in external stellar systems. Through mathematical analysis of the motions in the stellar system he argued that the spiral shapes of many nebulæ is derived from the fact that they are made up of rotating assemblies of stars. In such rotating systems there are sometimes perturbations and he argued that the spiral forms are a consequence of such perturbations. Stars that move in almost circular orbits can experience tidal effects that force the matter in the nebula to spiral forms. These tidal effects can occur when two stellar systems pass close to each other: the gravitation draws out spiral forms in the stellar systems. With close passages nebulæ can be attached more permanently to each other.[392] There is also a temporal aspect to the forms of the stellar systems. Spiral arms might disappear and the matter that is drawn out in spiral shapes can fall back in the system, he argued in 1930. The forms of the nebulæ, he argued later in 1934, are signs of an evolutionary track. The development is probably from an irregular object, over spiral nebulæ like M33, M51 and M81 to elliptical objects.[393]

Lindblad's theoretical work ran parallel with a remarkable achievement as scientific entrepreneur and administrator. He expanded the small group that he had around him in Uppsala to a larger group that came to play a dominating role in Swedish astronomy and he developed the international contacts. In this he built up a large observatory that became a new home for his group after he had managed to get the Stockholm professorship in 1927. The Royal Academy of Science was about to construct a new observatory, and Lindblad could as the new director influence the new observatory. By becoming professor just as the Academy was about to embark on the construction of a new observatory, Lindblad had quite considerable power in his hands to influence the future of astronomy. A newly appointed professor could of course in many ways direct the astronomical work, but if he also could direct the kinds of instruments present in the new observatory, he had freedom to pursue his project.

---

[392] Bertil Lindblad, "Kosmogoniska problem i samband med nyare föreställningar om stjärnsystemets natur", *PAT* vol 7 (1926), 125-134; Bertil Lindblad, "On the Dynamics of the Stellar System", *Arkiv* vol 21A no 3 (1928).

[393] Bertil Lindblad, "On the Development of Spiral Structure in Anagalactic Nebulæ", *Arkiv* vol 22A no 11 (1930); Bertil Lindblad, "On the Evolution of a Rotating System of Material Particles, with Applications to Saturn's Rings, the Planetary System and the Galaxy", *MNRAS* vol 94 (1934), 231-240.

A New Observatory

After the first world war, the construction of a new astronomical observatory was being discussed in Swedish astronomy. Some astronomers argued that Sweden needed a central observatory, more concentrated on research, but funding was a problem.[394]

Where was such an observatory to be placed? The choice of site depends on a wide range of various factors. As John Lankford has argued, the process of site selection before the construction of an astronomical observatory is an interesting field of inquiry for historians of astronomy.[395] The process has an environmental component, since the place ought to have optimal atmospheric conditions for astronomy, avoiding man-made as well as natural obstacles. (On the other hand, the reverse has been argued: observatories have at some instances been criticised because they were about to be constructed in environmentally sensitive areas.) The institutional context is also a factor: if the observatory mainly is to be used by students doing observations as part of their training or for popularisation of astronomy, then it is important to have it close to civilisation; if the main task is research, it is better to place the telescopes on a distant mountaintop. (This has another con-sequence for some astronomical observatories: astronomers have some-times been forced to spend periods of time together on an isolated mountaintop.) There might be astronomical reasons behind site selection, like when observatories have been placed in the southern hemisphere, on account of the southern sky being less studied than the northern. Technical and economical factors always put constraints on the construction of observatories: remote sites that are good atmospherically, could be too expensive. Sometimes, the funding agencies behind the resources have had reasons to argue for a special location. Political factors can play a part, for example when observatories are to be placed in foreign countries or be constructed as co-operation between countries. Colonial experience and colonial technologies have been assets for doing astronomy in places far away from Europe. In short, several factors can be thought of as playing a role in the process of site selection.

---

[394] Östen Bergstrand to Knut Lundmark September 6 1921, Lundmark's collection, LUB.

[395] John Lankford, *Americal Astronomy: Community, Careers, and Power, 1859-1940* (Chicago, 1997), 212 n.48.

This web of factors plays a part in the process leading to a new observatory. Charting this web can say something about the history of astronomy. It can put light on the question why a discovery or a work is made at a particular place. It can explain some parts of the institutes that have been the daily and nightly workspaces for astronomers. This web can explain parts of the development of what it means to be an astronomer, and what astronomers do in their work.

*

One possible placement for a new Swedish observatory was abroad. That would allow astronomers to observe the southern sky, and also seek out a climate more favourable to astronomical observations, than the Swedish climate.

Several Swedish astronomers thought about getting an observatory in the southern parts of the world. In 1922, Bertil Lindblad argued that his observatory at Uppsala ought to get a reflector of fairly good dimensions, placed somewhere in a southern location.[396] In an article published in *Populär astronomisk tidskrift* in 1923, Hugo von Zeipel posed the question: is Swedish astronomy internationally competitive?[397] He discussed the facilities at the Swedish observatories. The Lund staff of computers was seen as something that the other observatories ought to emulate. Otherwise resources were scarce: astronomy was "the step-child among the sciences in Sweden". The country lacked an observatory with science as its main purpose. Overburdened with teaching, not enough research was done at the observatories connected with the universities. Astronomy, regrettably, had no institutes like the National Museum of Natural History or the Nobel institutes that were soon to be founded in physics, chemistry and medicine (a Nobel institute already existed in physical chemistry).[398] The situation was not at all good for astronomy, von Zeipel

---

[396] Bertil Lindblad to Knut Lundmark, February 4, 1922, Lundmark's correspondence, LUB.

[397] Hugo von Zeipel, "Några önskemål för svensk astronomisk forskning", *PAT* vol 4 (1923), 62-64.

[398] Bernhard Hasselberg and Vilhelm Carlheim-Gyllensköld had made an attempt to organize a Nobel institute for astrophysics some years before von Zeipel's plea, but their attempt had failed. Robert Marc Friedman, "Nobelprisen och de vetenskapliga disciplinernas historia: Preliminära tankar och principer", in Gunnar Broberg, Gunnar Eriksson, Karin Johannisson eds., *Kunskapens trädgårdar: Om institutioner och institutionaliseringar i vetenskapen och livet* (Stockholm, 1988), 136-152, 148ff.

argued; in exactness it came close to pure mathematics, and the astrophysicists had the opportunity to study matter in its most extreme forms; nevertheless, this, perhaps the most noble of the sciences, had scarce resources.

Something had to be done. The proposal made by von Zeipel to promote "the most noble" science in Sweden was that the existing observatories were to be furnished with modern instrumentation where such was lacking, especially at Lund and Stockholm. Stockholm and Uppsala ought to get access to modern and rational computing bureaux like the one in Lund. Establishing a national observatory was the main part of von Zeipel's proposal. Such an institute was to be financed by the state and have a set of fully modern instruments. A reflecting telescope with one meter aperture and three meters focal length was what he had in mind, equipped for spectral analysis, nebular photography and photometry. It was to be joined by a long-focus refractor for precise measurements of the positions of heavenly bodies. Observations were to be handled at a bureau of computation and measurement. The institute was to be wholly focused on research. A small staff was to be running the institute; astronomers would come from their home institutes to the proposed national observatory for periods of time to do research. Another modern part of the proposal was the siting: it was to be placed abroad, von Zeipel argued. Climate in Sweden is not very good for astronomy. The northern sky having been studied quite extensively, it was seen as desirable to place the proposed observatory in the southern hemisphere, like in South Africa.

For some years, the question was being discussed among astronomers. Astronomers at Lund were discussing such a southern observatory. Walter Gyllenberg was interested in leaving theory altogether to concentrate on observational work. He thought about taking up a program in radial velocities of southern stars at such a proposed Swedish southern station.[399] Charlier discussed the matter with R.T.A. Innes at the Union observatory of South Africa. Charlier inquired about the possibilities of a Swedish station in South Africa. Innes replied that such possibilities indeed existed. The Swedes were welcome to mount their instruments on the grounds of the Union observatory.[400] Further on, in 1929, Harvard ob-

[399] Axel Corlin to Knut Lundmark March 24 1927, Lundmark's correspondence, LUB. Corlin had got this information after talking to Charlier and Gyllenberg.

[400] R.T.A. Innes to C.V.L. Charlier, March 30 1927, Charlier's correspondence, LUB.

servatory welcomed a co-operation with the Swedish astronomers; Harvard could help establishing a Swedish southern station, by sharing the knowledge gained when Harvard had begun operating their station at Bloemfontein in South Africa. Harvard had by then developed a working relationship with the local authorities.[401]

Another alternative was sending Swedish astronomers abroad to work as guest observers with larger telescopes and under southern skies. Knut Lundmark and Bertil Lindblad were active at Californian observatories for a couple of years early in their careers. Astronomers were sent to Germany and Egypt to procure plate material: for example Björn Svenonius, Erik Holmberg and Anders Reiz. Åke Wallenquist worked for some time at the Bosscha observatory at Lembang, in Java. He worked on the various projects the observatory was running, on variable stars; he also tried to collect observations for fellow Swedish astronomers. Lundmark wanted his help in getting observations of the Magellanic Clouds. Wallenquist's experiences were not altogether positive; he did not get along well with the observatory director whom he claimed did not know more than the very basic areas of astronomy. Wallenquist did not like the geography, far away from western civilisation and the company of fellow scientists. At times, he claimed to suffer very much.[402]

The plans for a Swedish observatory abroad were discussed but did not materialise during the period studied here. One reason was the relative ease of getting funds for a new observatory placed in Sweden, as will be discussed below. The main expansion in the inter-war period instead was domestic.

*

One of the first tasks Lindblad took up when he had become professor was to work on the new observatory. The question of a new observatory had been raised before Bohlin resigned, and a committee had been working on the question; in April 1927 it had issued a preliminary report.[403] A refurbishing of the present observatory on *observatoriekullen* in the cen-

[401] W.J. Luyten to Knut Lundmark February 26 1929, Lundmark's correspondence, LUB.

[402] Åke Wallenquist to Knut Lundmark September 30 1929, July 16 1932, July 30 1932, November 20 1932, Lundmark's correspondence, LUB.

[403] The observatory committee's preliminary report, 8 April 1927, Minutes appendices of the Royal Academy of Sciences 1927 no 414, KVA.

tre of Stockholm would not be practical. An observatory had been present there since 1753 and had suffered more and more from light-pollution as the city of Stockholm grew in size. It would result in a second-rate observatory, because of the site. A new construction at a better site was projected to cost about 3 million kr. One way of raising such amounts of money was to sell the rights to use *observatoriekullen* to the city of Stockholm. (The city already owned the site but the Academy had the right to use it (*dispositionsrätten*).) A proposal was sent off to the city.[404]

The city of Stockholm agreed to buy the rights to use the observatory hill for 900 000 kr, and the Academy also managed to get 1 million kr from the Knut and Alice Wallenberg foundation for "the construction of a modern observatory on the so-called Karlsbader mountain at Saltsjöbaden on the condition that this observatory was called the Stockholm observatory".[405] The plans for the observatory included four main instruments: a reflector with 1 meter aperture and 3 meters focal distance, a double refractor of 0.5 meter aperture, a double astrograph with 0.3 meter diameter, and a meridian circle. The estimated costs (in June 1928) ran to 2.1 million kr.[406]

In a report, Bertil Lindblad discussed the site selection of the new observatory.[407] He noted that when the observatory committee had started work in 1927 the Karlsbader hill was considered a candidate. More research had confirmed this choice: the Karlsbader hill had free sight, no growth of industries was planned in the vicinity, meteorological conditions were quite good and there were good communications with Stockholm.

It is difficult to decide in what way the wishes of the Wallenberg foundation played a part the site selection process. In the Academy's minutes it is stated that Lindblad did not know of "the donor's regulations when it comes to the place where the observatory was to be built" when Lindblad drew up the plan for the observatory.[408] On the other hand, there are indications that Knut Wallenberg demanded that the observatory should be placed on the Karlsbader hill in Saltsjöbaden. Even if

---

[404] §14 Minutes of the Royal Academy of Science 13 April 1927, KVA.

[405] §2-4 Minutes of the Royal Academy of Science 20 June 1928, KVA.

[406] "P.M. Kostnadsberäkning för ett nytt Stockholmsobservatorium", dated 19 June 1928, Minutes appendices of the Royal Academy of Sciences 1928 no 572, KVA.

[407] Lindblad's report on the new observatory's placement dated 20 June 1928, Minutes appendices of the Royal Academy of Sciences 1928 no 571, KVA.

[408] §2-4 Minutes of the Royal Academy of Science 20 June 1928, KVA.

scientific reasons were important when it came to select the new site for the observatory, it is hard to ignore that Knut Wallenberg wanted to have the observatory located in Saltsjöbaden. The wish of the donor probably meant something in this case.[409]

The suburb of Saltsjöbaden was very much the work of Knut Wallenberg and a product of the 1890's. It was constructed with the ideal of providing a place of natural beauty that was still close to Stockholm. Saltsjöbaden was situated between the capital and the beautiful Stockholm archipelago. The wealthy banker Knut Wallenberg had bought 1600 acres of land by the sea. In August 1892 the king's ship Drott arrived at Saltsjöbaden and Oscar II put the founding stone in place in what was to become a luxurious hotel. Many villas were subsequently constructed and sold. Saltsjöbaden catered to a group of people that wanted to get out of the hectic life in urban Stockholm, while still be so close to the capital that they could pursue their careers there. It was hailed as a place of natural beauty, for relaxation and recreation.[410]

Wallenberg constructed a railway that connected Saltsjöbaden to Stockholm. His railway company Järnvägs AB Stockholm-Saltsjöbaden, was a central actor in the development of the idyllic suburb to Stockholm. The company developed villas that were later sold; it was also involved as contractor when the observatory was built, by constructing a road that led up to the Karlsbader hill. Wallenberg thus had many connections and interests in Saltsjöbaden. This is one background to his decision to fund the observatory in Saltsjöbaden. Perhaps he thought that the addition of the nation's premier astronomical observatory would draw status to the suburb of Saltsjöbaden. Wallenberg had a history of patronage; he had funded parts of the opera house in Stockholm. Whatever the reasons for his activities as donor to culture, in this case the observatory fell in the same geographical location as the place where he was developing the new suburb.

The observatory's architecture is quite remarkable. On top of the main building there is a large dome that shelters the double refractor. The refractor was to be the longest of the instruments, therefore its dome was also the largest. By placing the dome for the refractor on top of the main building, that spot was highlighted.

---

[409] Gunnar Hoppe, *Till landets gagn: Knut och Alice Wallenbergs stiftelse 1917-1992* (Stockholm, 1993), 178ff. Ilse-Beth Byström, "Stockholms observatorium i Saltsjöbaden: Ett byggnadsverk av Axel Anderberg", *AT* vol 15, 152-163, 153f.

[410] Staffan Tjerneld, *Wallenbergs* (Stockholm, 1969; 1971), 33-43.

Axel Anderberg, the observatory's architect, had been the architect for the Royal Academy of Sciences since 1923. In the beginning of his career he had designed the Stockholm opera house (built 1891-1898 and also funded in part by Wallenberg), then he worked with churches until about 1910, and afterwards several scientific buildings. He died in 1937.[411]

Anderberg had earlier in buildings for the Royal Academy of Science and the Museum of Natural History in Frescati, Stockholm, (1915) shown that he could make monumental buildings for science. The observatory in Saltsjöbaden had such a monumental look. The observatory is easily spotted from far away; it is conspicuous because of the "monumental character" of its main building that was Anderberg's intention.[412] Now the observatory had a building that was suitable for the role it was to play: the new, powerful, centre of astronomy in Sweden with backing by a leading economical force – the Wallenberg family – and the Royal Academy of Sciences.

*

Lindblad's priority was to get a large astrograph. Such an instrument would make it possible for him to continue the work in wide-field spectroscopic observations he had started at Uppsala. A larger astrograph than the one at Uppsala would be a natural choice in developing the programme. Second on the list of priorities was a reflector in the one meter class. Third, he wanted a refractor with long focal length for precise astrometric measurements, but only if there were sufficient funds, otherwise this instrument would be left out. Of least importance was a meridian circle.[413]

The observatory ordered a reflector with a one meter aperture from the instrument firm Howard Grubb of Dublin. Grubb also supplied a double refractor with 50 and 60 centimeters aperture. From Carl Zeiss an astrograph with an aperture of 40 centimeters was ordered. These three types of telescopes were suitable for different ways of observing the sky. The refractor, suitable for precision measurements of stellar positions, belonged to classical astronomy. The reflector could be used for studies of the spectra of faint stars, or spectroscopy in a higher dispersion. The

---

[411] Byström, "Stockholms observatorium", 163.

[412] Axel Anderberg, "Stockholms observatorium: Beskrivning av byggnaderna", 2, Bertil Lindblad's papers, F7B:1, KVA.

[413] Bertil Lindblad to Knut Lundmark 15 August 1928, Lundmark's papers, LUB.

astrograph was a survey-type instrument that could register the spectra or colours of many stars on a single plate. In size, it was bigger than the astrograph used at Uppsala.

Besides the telescopes there was auxiliary instrumentation. Spectroscopes with various resolutions were available for the reflector. For the astrograph a large objective prism was available. An important auxiliary instrument were photometers that measured intensities on the photographic plates, and thus made possible a quantitative analysis of the brightness of astronomical objects. One photometer was used for measuring the brightness of stars. Another photometer made it possible to make microphotometric tracings on the plates, hence getting diagrams of spectra.

The Saltsjöbaden equipment was diversified. There was a long-focus refractor devoted to photographic astrometry, a large reflector and a large astrograph; each instrument catered to one part of astronomy and astrophysics. While the Lund refractor with the addition of the spectroscope and the Uppsala double refractor were instruments that combined several areas of work, there was less compromising going on in Saltsjöbaden. The instruments were more specialised, and most of them devoted to astrophysics. The instrumental diversity also reflects the high level of funding available to Lindblad and the Saltsjöbaden project.

With the observatory at Saltsjöbaden, a number of new positions became available in Swedish astronomy. Young astronomers were hired to work at the observatory, and the new observatory was seen as attractive by many astronomers. Some began to think that Saltsjöbaden was the only place in Sweden where astronomy could be pursued properly. A number of new positions became available at the new observatory. One was the position as observator, for which Lindblad suggested Gunnar Malmquist. (Knut Lundmark had tried to get Östen Bergstrand and Bertil Lindblad to propose Walter Gyllenberg as observator at Saltsjöbaden. Lundmark wanted to get rid of Gyllenberg, who was a proud person and easily could feel neglected if he didn't get the post at Stockholm, he argued. One can suspect that Lundmark would be happy to get Gyllenberg out of the way in Lund.[414]) Malmquist's style of astronomy was a mixture

[414] Lundmark to Lindblad 31 October 1930, Lindblad's papers F13:2, KVA; Lundmark to Östen Bergstrand 6 April 1930, Lindblad's papers F13:1, KVA; Bergstrand to Lindblad 7 April 1930, Lindblad's papers F13:1, KVA. The discussions between Lindblad and Bergstrand shows that Bergstrand and Lindblad collaborated on matters of astronomical politics in the academy.

of theory and observation. He had worked with theory under Charlier, but also observed the colours and magnitudes of faint stars. Lindblad paid attention to Malmquist's mixture of theoretical studies of the distribution of stars and observations of astrophysical data in the report to the astronomy election committee of the Academy.[415] Malmquist's empirical work in astrophysics made him interesting to the group around Lindblad. At the observatory in Hamburg he had made photographic observations of stellar colours with a telescope of the same dimensions as the one at Saltsjöbaden.[416]

Malmquist had started out in stellar statistics but gradually moved to a style that was conceived by Lindblad and others as more modern. Malmquist had even been critical of Charlier's stellar statistics. In a paper in 1926 he had discussed the methods of the stellar statistical method by a simulation. He studied a fictitious stellar system with a clear pattern: in the centre there was a globular cluster of stars; next a region without stars, and then another region with stars. The space density of stars was put to be the same. By counting backwards he tried to see what result the stellar statistical method would have. Malmquist envisioned an observer in the middle of the system. Given the model and the methods used in stellar statistics, the result was a picture that did not resemble the model: the inner group of stars was hardly visible in the results, whereas the empty space was filled with stars; gradually the density shrank after that, rather than rising from zero. The statistical methods were far too simple for getting a proper picture of the universe.[417]

Malmquist took up the post as observator on 1 January 1931 and then got access to the kind of instruments that were not present at his former

[415] Lindblad's report to the astronomy election committee of the Royal Academy of Sciences, dated 28 October 1930, Bertil Lindblad's papers, F7B:1, KVA; §6, minutes of the Royal Academy of Sciences 3 December 1930, KVA.

[416] Gunnar Malmquist, "Bestämning av stjärnornas färg i större skala med Seares' metod", *PAT* vol 5 (1924), 19-25; Gunnar Malmquist, "Hamburgs observatorium i Bergedorf", *PAT* vol 6 (1925), 35-43.

[417] Gunnar Malmquist, "Einige Bemerkungen über die räumliche Verteilung der Sterne", *Vierteljahrsschrift der Astronomischen Gesellschaft* vol 61 (1926), 244-250. It has been argued that problems inherent in the stellar statistical methods demonstrated by Malmquist's simulation were the reason for the ultimate lack of success for stellar statistics, rather than the school's neglect of interstellar absorption, as has sometimes been proposed. Hans Kienle, "Historical Development of our Ideas Concerning the Structure of the Galaxy", in L.N. Mavridis ed., *Structure and Evolution of the Galaxy* (Dordrecht, 1971), 1-16, 11f.

post at the Lund observatory. He started a programme of observations on the colours, magnitudes and spectra of stars in the vicinity of the north galactic pole. He worked on devising methods for measuring colour indices of stars.[418] Malmquist also ran a large observational programme that measured the spectra of some 2800 stars at high galactic latitudes. The stars, down to about magnitude 13.5, were observed with the Saltsjöbaden astrograph fitted with an objective prism. This big survey-type instrument was directly geared towards the kind of studies developed by Lindblad. Malmquist measured the spectra with a self-registering photoelectirc microphotometer, an instrument that gave tracings of the spectra, diagrams that gave astronomers another way of measuring features in the spectra, compared to visually inspecting spectra on the plates. Malmquist studied some 11 000 tracings for his projects of classifying the stars' spectra. He also used Lindblad's criteria for differentiate between stars of different luminosity classes.[419]

Yngve Öhman worked on continuing the methods of spectroscopy that Lindblad and Bergstrand had developed. He was first connected to the Uppsala observatory, but then moved to a position at Saltsjöbaden when that observatory was constructed. Öhman began working on methods of microphotometric tracings of stellar spectra. Prior to his move to Saltsjöbaden, he had used photometers at the Uppsala physics department and observatory on photographic stellar spectra.[420] Öhman developed Lindblad's methods even further. Öhman measured the density on the photographic plates with a photometer instead of estimating the intensity of spectral lines, thus aiming for a higher precision.

One central way of dealing with spectra at Saltsjöbaden was to analyse them by way of measuring the density of the photographic plates. Diagrams of the distribution of energy in the spectra were the product of this method. Classification of spectra had traditionally been an activity where astronomers looked at the spectra (either directly through a spectroscope, or on photographic plates); that was the way Dunér had worked, and that was also the way some of Lindblad's contemporaries

418 Gunnar Malmquist, "On the Determination of Colour Indices with Reflectors", *Stock. Ann.* vol 13 no 1 (1938).

419 Gunnar Malmquist, [no title], *Transactions of the International Astronomical Union* vol 6 (Cambridge, 1939), 459-461.

420 Yngve Öhman, "Photographic Density-Diagrams of Short Stellar Spectra", *Arkiv* vol 19B no 6 (1926), Yngve Öhman, "Photometric Studies of Effects of Luminosity and Colour in Short Stellar Spectra", *Arkiv* vol 20A no 23 (1927).

worked. Galison has argued that this qualitative way of working, which re-introduced the observer's judgment as an important element, supplanted a mechanical visual ideal in science that had been in vogue from the 1830's to about the 1920's, that was more quantitative than qualitative. Galison use, as an example of this trend, the classification of stellar spectra done in the 1940's by W.W. Morgan, Philip C. Keenan and Edith Kellman.[421] But this qualitative method that hinged on the interpretation by the astronomer was supplemented with a trend towards measurement on plates. Astronomers increasingly used optoelectronic systems for producing tracings of the distribution of energy throughout the spectrum. At the Uppsala and later Saltsjöbaden observatory, a way of working with spectra was developed that involved the reduction and analysis of spectral plates after they had been observed. These reductions were not about the astronomer 'judging' which spectral type a particular star had by looking at the plates. First, the plates were measured with a photometer, often a kind of instrument that automatically produced tracings of the distribution of intensity throughout the spectrum. This quantitative data in a diagrammatic form was then used for determining the spectral type or what type of luminosity class the star had. The "registrograms" were measured at several points in the spectrum, which gave the strength of the cyanogen absorption (in late-type spectra) or Balmer lines (early type).[422] The astronomers got quantitative indices of spectral features that could be used for determination of stellar luminosities and, hence, distances (after a calibration of the criteria by observing stars of known distances).

This type of photographic spectral photometry was at the heart of the stellar astronomy group discussed here and represents a trend that was the very opposite of what Galison claims.[423] Galison's use of the MKK atlas is not suitable for what he want to show. The work of Morgan, Keenan, and Kellman represents the continuation of a tradition in spectral classification. Many astronomers between the 1860's and the 1920's classified spectra by judging the appearance of spectral details; that was how Angelo Secchi, Lewis Rutherford, Hermann Carl Vogel, Nils

---

[421] Peter Galison, "Judgment against Objectivity", in Caroline A. Jones & Peter Galison, *Picturing Science, Producing Art* (New York & London, 1998), 327-359.

[422] Bertil Lindblad & Erik Stenquist, *On the Spectrophotometric Criteria of Stellar Luminosity, Astronomiska iakttagelser och undersökningar å Stockholms observatorium* vol 11 no 12 (1934), 8ff.

[423] Hearnshaw, *Analysis*, chs 4&5.

Dunér, Antonia Maury, Williamina Fleming, Annie Jump Cannon and others went about classifying stellar spectra.

In his PhD dissertation, Öhman studied the spectra of 940 stars on plates exposed with the Uppsala astrograph and an objective prism.[424] He used a photometer to measure the absorption of light in the spectral lines. For most of the stars, absolute magnitudes were determined from measurements of hydrogen line intensities and colours.

Öhman also worked on spectra with higher resolution, and developed the methods of measuring spectra on plates further. Rather than measure the density of the photographic plates at several points, instruments were constructed that were self-registering; they made tracings of the intensity in the spectra. During a visit to Mount Wilson observatory in 1933 Öhman made a discovery: for the first time spectral lines from metal hydrates were observed in red dwarf stars. He worked here on a programme with Merrill and Humason that studied the red part of stellar spectra. Öhman also found calcium hydride in several red dwarfs, the first time that substance had been observed in any star. He also showed that these hydrides made good luminosity criteria: the bands were faint or invisible in giant stars.[425] Some of these studies, containing a number of firsts or 'discov-eries' were possible to make using the Saltsjöbaden instrumentation, thus for a short while making real the dream of a modern Swedish astronomical observatory performing front-line observations.

Öhman made the first tracings of white dwarf spectra. He attributed the broad spectral lines to the Stark effect (spectral lines are broadened if they are produced under high pressure). The pressure on a white dwarf star was very high, according to his measurements.[426] He also started work on the polarisation of celestial objects. In 1934 he discovered, using a polarimetric equipment on the Stockholm reflector, that the light from

---

[424] Yngve Öhman, *Spectrophotometric Studies of B, A and F Type Stars* (*Nova Acta Regiæ Societatis Scientiarum Upsaliensis* ser 4 vol 7 no 3) (Uppsala, 1930).

[425] Yngve Öhman, "Spectrographic Studies in the Red", *ApJ* vol 80 (1934), 171-180; Yngve Öhman, "The Red Spectral Region of Dwarf Stars of Class M", *Stock. Ann.* vol 12 no 3 (1936); Yngve Öhman, "On the Bands of Magnesium Hydride in Stellar Spectra", *Stock. Ann.* vol 12 no 8 (1936). Cf. John Hearnshaw, *Analysis of Starlight*, 309f.

[426] Yngve Öhman, "Note on the Spectrum of a Probable White Dwarf in h Persei", *MNRAS* vol 92 (1931), 71-72. Yngve Öhman, *Some Preliminary Results from a Study of Hydrogen Absorption for Stars in h and χ Persei*, *Stock. Ann.* vol 12 no 12 (1935), 16. Cf Hearnshaw, *The Analysis of Starlight*, 352-360.

the variable star ß Lyræ showed effects of polarisation.[427] Öhman was interested in optics and new types of astronomical instrumentation. He reported on Bernhard Schmidt's new telescope type in *Populär astronomisk tidskrift*.[428] Several subsequent investigations by Öhman were centred on measurements of the polarisation of light from celestial objects. In 1938 he developed a new kind of instrument for studies of the sun. Based on the polarisation of light, it was a filter that made it possible to observe the sun in a very tight span of wavelength. The astronomer could therefore observe solar phenomena like protuberances and flares with more simplicity than the earlier methods (the spectroheliograph deviced by Hale). The type of instrument has since become very much used in solar astronomy.[429]

Most astronomers at Saltsjöbaden were active in areas that belonged to Lindblad's (and Bergstrand's) spheres of interest. A mixture of observational and theoretical astrophysics, a theoretically informed observational astrophysics was the style of work at Saltsjöbaden. But the observatory had one large instrument that was designed for classical astronomy, the double refractor. To operate that instrument, Sten Asklöf was recruited. He was more or less the only young astronomer that had had any extensive experience with photographic astrometry with long focus refractors.

Asklöf had received a PhD at Uppsala on a thesis on parallax determinations with the 33/36 cm double refractor there. During a period of time he had worked with a large astrometrical instrument at Leander McCormick observatory. There was, however, one problem of a personal nature. Asklöf was not on perfectly friendly terms with a couple of members of the stellar astronomy group, Schalén and Öhman, during his

---

[427] Yngve Öhman, "Effects of Polarisation in the Spectrum of ß Lyræ", *Nature* vol 134 (1934) 534-536. It was, Öhman claimed, too early to speculate on the reason for the polarisation. One possible reason was a star with aspherical shape.

[428] Yngve Öhman, "En kombination av reflektor och refraktor", *PAT* vol 12 (1931) 161f.

[429] Yngve Öhman, "A New Monochromator", *Nature* vol 141 (1938), 157-160. The construction had been published in 1933 by Bernhard Lyot in a paper that Öhman claimed not to know about. A difference between Lyot's and Öhman's work was that Öhman used the new material Polaroid film in his version of the instrument. Also, Lyot had not been able to construct a working model of the instrument. Yngve Öhman, "A New Monochromator", *Nature* vol 141 (1938), 291.

time at Uppsala. The reason for this animosity is unclear, and it is uncertain if it led to any problems during Asklöf's years on Karlsbader hill.[430]

*

When the observatory was to be moved from Observatory Hill to Saltsjöbaden, the organisation of the observatory was changed. Instructions telling the assistant astronomers what to do were changed, and the people on these assistant posts were exchanged.

It was emphasised that the assistants were to be selected on scientific merits, since they were to do scientific work. A limit of three years was also introduced for these posts. Lindblad got more opportunities to make the institute work according to his plans by these changes: he could put the right kind of people on the posts.[431]

The new situation, which called for a more active participation in scientific work by the assistants, was something quite different from the prior situation. Lindblad argued that these posts should be intended for younger scientists in the beginning of a scientific career, which wasn't the case with the two assistants holding the posts, Ansgar Roth and A. Lundborg. "During their long time of employment they [Roth and Lundborg] have hardly shown any signs of independent scientific work", Lindblad noted in his proposal to the Academy. During the two years since Lindblad had taken up leadership of the observatory they had shown no interest in research. They, instead, wasted their time on extramural activities, Roth as a science writer in the press and Lundborg as a teacher in a secondary school. The academy decided to kick out Roth and Lundborg, which meant that they lost both income and a place to live (they had rooms at the old observatory).[432]

Roth and Lundborg answered with a report where they claimed that Lindblad's accusations were false.[433] Old age, which Lindblad had criti-

---

[430] See, for example, Sten Asklöf to Knut Lundmark 18 January, 18 April, 9 June 1929, 9 April 1930, Lundmark's papers, LUB.

[431] §4, "Förslag till ändring av §12 mom. b och c samt av §18 i Akademiens grundstadgar", Minutes appendices of the Royal Academy of Sciences 1930, KVA.

[432] Lindblad's proposal dated 28 November 1929, Minutes appendices of the Royal Academy of Sciences 1930, no 949, KVA; §22 Minutes of the Royal Academy of Sciences 8 January 1930, KVA.

[433] Ansgar Roth and A. Lundborg to the Royal academy of sciences, 16 May 1930, Bertil Lindblad's papers F7B:1, KVA.

cized them for, was stated by them to be a merit, and the lack of independent research work they blamed on the conditions at the observatory. Such work was impossible under the previous leader Karl Bohlin:

> That reign of terror that was Bohlin's thirty years as leader of the Stockholm observatory is impossible to describe; only those who have lived at the observatory could grasp its magnitude. The observatory was in decay; the rooms lacked proper heating or cleaning; books were suffering in conditions of high humidity, the instruments were attacked by corrosion; hysterical outbursts of the loudest [*larmande*] type were occuring daily – and nightly –; the leader's maniacal views and sick suspicious mind towards everything and everybody was daily at a grotesque level and gave a permanent feeling of unpleasantness and anxiety in the surroundings; the milieu was, in short, the very opposite of that what is necessary for productive scientific work.

The words are harsh and can of course be attributed to the immense pressure Roth and Lundborg felt; they were, after all, losing their jobs and their apartments. But there are other indications that Bohlin was a problematic person to work under (see earlier chapter on Lundmark). Bohlin's problematical personality made the working situation very hard. Conditions at the old Stockholm observatory had been "embarrasing – not to say detrimental"; such explanation for the lack of research work by the staff under Bohlin could probably have some amount of truth. Also, they argued, routine tasks like calculating data for the almanac had taken up much time. (The Royal Academy of Sciences had a monopoly on almanac publishing, and it was handled by the Academy's observatory, which meant that the assistant astronomers had to perform a lot of calculations in preparation for the almanac.)

Lindblad had, of course, no problems in defending himself.[434] He argued that the assistants had complained too much. He did not comment on the conditions during the reign of his predecessor, but commented solely what had happened during the two years that he had been director of the observatory. Roth had shown a direct unwillingness when it came to perform observations: a programme of observations of variable stars and of spectroscopy with the old reflector had led to nothing. Lindblad also commented on Roth's slow work on calculations. He also hinted that such old gentlemen would have problems performing the demanding observational work that would be a daily (or nightly) routine at the new observatory. In this discourse, observational work was described as some-

[434] Bertil Lindblad to the Royal Academy of Sciences dated 27 May 1930, Minutes appendices of the Royal Academy of Sciences 1930 no 527, KVA.

thing that demanded youthful strength, agility and stamina. Roth and Lundborg simply weren't fit enough for such work.

The results of this was that new people got posts at the observatory. The first assistant hired was Yngve Öhman, then still without a PhD (he had a licentiate degree) and assistant at the Uppsala observatory. In his arguments for Öhman, Lindblad described him as independent and a producer of high quality work with skills in "one of modern astronomy's most important supporting disciplines: physical optics". He was, according to Lindblad, a good observer that would be able to make good use of the new instruments at Saltsjöbaden.[435]

Saltsjöbaden was hiring young people. Such was the strength of the new observatory that Knut Lundmark feared that it would make life difficult for the university observatories. Several astronomers had told him that Saltsjöbaden would be the only place were astronomy could be done properly in Sweden.[436]

When Gunnar Malmquist was preparing to leave Lund for Saltsjöbaden, two younger Lund astronomers, H. Nordström and Jöran Ramberg, made inquiries about Saltsjöbaden observatory. Malmquist wrote to Lindblad about them and told him about their wishes, their merits and personalities; the communication was "confidential, they do not wish it to be known that they want to get away from Lund".[437] He noted that Ramberg, besides his scientific merits, also had skills applicable to instrumentation, as he had worked as an apprentice in a watch repair shop. Ramberg's wish became reality and he moved to Saltsjöbaden at the end of 1932. By that time, Lundmark was in the US. Ramberg wrote Lundmark that he wanted to get to Saltsjöbaden because that observatory had considerably better resources for research. Gyllenberg also wrote Lundmark about Ramberg's move, saying that it was a pity since Ramberg was one of the brightest of the young scientists at Lund.[438] From his temporary vantage point in the US, Lundmark could only watch as one of the promising young astronomers left his institution. Gyllenberg's let-

---

[435] Lindblad to the Academy 5 May 1930, Minutes appendices of the Royal Academy of Sciences 1930 no 433, KVA.

[436] Knut Lundmark to Yngve Öhman 2 March 1935, Lundmark's papers, LUB.

[437] Gunnar Malmquist to Bertil Lindblad 24 December 1930, Lindblad's papers, F13:2, KVA.

[438] Jöran Ramberg to Knut Lundmark 24 November 1932; Walter Gyllenberg to Knut Lundmark 29 November 1932, both in Lundmark's papers, LUB.

ter probably only made him feel worse, pointing out the attraction of the new observatory in Saltsjöbaden felt by young astronomers.

The Milky Way over Uppsala

Stellar astronomy at Uppsala and at Saltsjöbaden had many points of contact. After Lindblad left for Stockholm, activities continued along the lines that Lindblad and Bergstrand had worked on.[439] Most Uppsala astronomers took part in the observational work that studied spectra and colours of large quantities of stars, with an ultimate goal of elucidating structures in the Milky Way system of stars. A local culture and tradition of observational astronomy and astrophysics had been strong in Uppsala at least since Nils Dunér became professor there in 1889. Dunér had started up observational astrophysics at Uppsala, Bergstrand had taken up such work around 1907, Lindblad had followed up with a development of new methods, and he built up a group of astronomers doing that kind of astronomy. Bergstrand was an important figure in promoting this kind of astronomy, rather than von Zeipel.[440]

Just like at Saltsjöbaden, efforts were made to enable astronomers to work quantitatively in analysing the photographic plates, by measuring the plates with a microphotometer to determine the brightness of stars. Spectral plates were analysed using registering microphotometers, an instrument that produced diagrams of the density across the spectrum. These diagrams were later measured and analysed to give information on intensities in various details in the spectrum. By developing the use of the microphotometer, astronomers wanted to move further away from the way of classifying spectra that dealt with visual inspection by quantifying the spectral information.

---

[439] At Uppsala Östen Bergstrand had a colleague as professor as astronomy, the theoretically inclined Hugo von Zeipel. (The post of *observator* had temporarily been made into a personal professor's chair for von Zeipel.) For example, he did work on the stellar energy problem, and he taught courses on modern astrophysics. A research group did not form around him.

[440] Bergstrand and von Zeipel were, for a period of about ten years, not on very friendly terms. Östen Bergstrand to Carl Schalén 5 September 1930, Schalén's papers, LUB; Axel Corlin to Knut Lundmark 15, 18, and 20 February 1921, Lundmark's papers, LUB; Knut Lundmark to Elis Strömgren 16 October 1921, Strömgren's papers, LUB; Östen Bergstrand to Elis Strömgren, 7 November 1927, Strömgren's papers, LUB.

At Uppsala, there was not a lack of theoretical interest altogether. The work was not purely observational in a fact-collecting way. One example of this mixture of theoretical and observational astrophysics is Carl Schalén. He arrived at Uppsala in the autumn of 1924, where he finished his PhD thesis in 1928.[441]

Schaléns first work was made in direct cooperation with Bertil Lindblad. In a paper in 1925 he studied the spectra of faint stars in the Milky Way in Aquila. In a series of early works, of which the PhD thesis was one, he determined distances and other data for stars with the spectroscopic techniques developed by Lindblad. Schalén was being schooled in to the local culture of scientific practice.[442]

Schalén spent a year in the US in the mid-1930's, visiting among other places Harvard. He developed a net of contacts with several American astronomers that worked on spectral classification, theoretical and observational studies of dark nebulæ, and other kinds of Milky Way astronomy.[443]

Schalén early on took up an interest in the question of interstellar absorption. Is there a general absorption of light as it travels through interstellar space, or is space transparent? The existence of localised clouds of obscuring matter had been known earlier; Schalén wanted to look for a general absorption. In a paper published in 1929, the result was that there is a general absorption in space that amounts to 0.5 magnitudes per kiloparsec.[444] Stellar statisticians had earlier discussed the possibility of a gen-

---

[441] Åke Wallenquist, "Schalén, Carl Adam Wilhelm", *SMoK*; "Schalén, Carl Adam Wilhelm", *Lunds universitets matrikel 1967-1968.*

[442] Carl Schalén, "Spectral Classification of Faint Milky Way Stars in Aquila", *Arkiv* vol 18 no 36 (1925); Carl Schalen, "Spectro-photometric Determinations of Absolute magnitudes of B and A Type Stars", *Arkiv* vol 19A no 33 (1926): Carl Schalén and Bertil Lindblad, "The Luminosities, Individual Parallaxes and Motions of B and A Type Stars", *Arkiv* vol 20A no 7 (1927); Carl Schalén, "The Space Distribution of B and A Type Stars in Bright and Dark Galactic Regions. Researches Based on Determinations of Spectral Type and Luminosity Class for 4300 Stars", *KVAH* ser 3 vol 6 no 6 (1928).

[443] See, for example, Carl Schalén to Annie Jump Cannon October 27th 1934, Annie Jump Cannon to Schalén 9 November 1934; Bart and Priscilla Bok to Schalen no date (end of 1934); Bart Bok to Schalén 15 February 1940; Jesse Greenstein to Schalén 25 February no year (= 1937); Schalén to Greenstein 20 March 1937 – all in Schalén's papers, LUB.

[444] Carl Schalén, "Om en allmän absorption av ljuset i världsrymden", *Beretn. om det 18. Skand. Naturforskermöde i Köbenhavn* (Köpenhamn, 1929); Carl Schalén, "Zur Frage einer allgemeinen Absorption des Lichtes im Weltraum", *AN* no 5656 vol 236

eral absorption of light in space, but many had dismissed the idea. Schalén's results came at about the same time as other astronomers found similar results, among others Bart Bok and Robert Trümpler. Trümpler, working at the Lick observatory, published a study in 1930 that showed a general absorption of 0.7 magnitudes per kiloparsec.[445] His results were thus about the same as Schalén's, who tried to tell him about the earlier paper; the answer was that he had not known of the result when he wrote the paper, otherwise he would have included a reference to it.[446]

Schalén also studied the properties of interstellar matter theoretically. He started using the theory of Mie in the early 1930's for studies of how light diffused when it passed through interstellar matter. By applying this theory to observations, it was possible to give an estimate of the size of particles in the interstellar matter, which turned out to be about 10^{-7} m.[447]

Schalén's work showed a mixture of theory and observations that was typical for the Swedish stellar astronomy school. He worked on observational programs that were rooted in the Lindblad-Uppsala way of doing astronomy.

Several other astronomers worked in stellar astronomy during the 1930's. Åke Wallenquist discussed the distances to dark nebulæ and studied open clusters photometrically.[448] Erik Stenquist used photographic material observed both at Uppsala and Saltsjöbaden to measure the magnitudes of stars, by measurement on the plates with a photometer. Stenquist also measured thousands of spectra, making tracings of the densities of the photographic plates. On these diagrams he measured the intensities of Balmer (hydrogen) spectral lines and other indicators of stellar luminosities. For these measurements on the spectra, he used the registrating

---

(1929), 249-258; Carl Schalén, "Om förekomsten av mörk materie i världsrymden", *PAT* vol 11 (1930), 22-44. Cf Bart J. Bok, *The Distribution of the Stars in Space* (Chicago, 1937).

[445] Erich Robert Paul, *Statistical Cosmology*, 228ff.

[446] Carl Schalén to Robert Trümpler 9 May 1930, Trümpler to Schalén 28 May 1930, Schalén's papers, LUB.

[447] Carl Schalén, "Untersuchungen über Dunkelnebel", *KVAH* ser 3 vol 13 no 2 (1934); Carl Schalén, "Über Probleme der interstellaren Absorption", *Nova acta R. Soc. Scient. Ups.* Ser 4 vol 10 (1936).

[448] Åke Wallenquist, "A Photometric Research on the Galactic Cluster Messier 11 (N.G.C. 6705)", *Nova Acta Regiæ Societatis Scientiarum Upsaliensis* ser 4 vol 10 no 4 (Uppsala, 1936); Åke Wallenquist, "Some Preliminary Results on the Space Distribution of White Stars in the Direction of the Galactic Centre", *Arkiv* vol 25A no 27 (1937); Åke Wallenquist, "On the Dark Nebula at theta Ophiuchi", *Arkiv* vol 25B no 28 (1937).

microphotometer at the Saltsjöbaden – somewhat later, in 1935, Uppsala observatory itself procured such an instrument.[449]

At the end of this time period, Gunnar Malmquist came to Uppsala as professor, from Lund by way of a post as observator at Saltsjöbaden. He was then firmly rooted in the style of stellar astronomy of Lindblad.

Peripheral Lund and the Stockholm-Uppsala Axis

The observatory at Saltsjöbaden was large and well-financed. It had by far the most modern astronomical and astrophysical instruments that were available in Sweden. Its inauguration also led to something of an expansion of astronomy in Sweden; more positions became available, especially for young astronomers who did observational astrophysics and stellar astronomy.

At Uppsala, Östen Bergstrand was still professor of astronomy. He played an important role in this process. He was mentor to both Lundmark and Lindblad, and it was at Bergstrand's observatory that Lindblad further developed the observational astrophysical methods that Bergstrand had begun. At Uppsala, Lindblad could early form a group of astronomers who worked under him on extending the Bergstrand way of doing astronomy and astrophysics. Bergstrand and Lindblad also collaborated in various ways in the Royal Academy of Sciences on furthering astronomy. Astronomers at Saltsjöbaden and Uppsala sometimes presented their work as being connected. It is thus possible to speak of something of an Uppsala-Stockholm axis in astronomy and astrophysics, at least during the 1930's.[450] Lund and Lundmark became more of an institute on the margin, led by an astronomer that felt himself to be more and more an outsider. One obvious example of this situation is the fact that Lundmark never took place in the Academy; other examples are statements by Lundmark in letters (see chapter on Lundmark).

---

[449] Uppsala yearly reports, *UUÅ 1934-35*, *1935-36*; Erik Stenquist, *A Spectrophotometric Study of the Cambridge Proper Motion Stars* (*Nova Acta Regiæ Societatis Scientiarum Upsaliensis* ser 4 vol 10 no 7) (Uppsala, 1936).

[450] Remarks to that effect are made in for example Bertil Lindblad, "Spectral Types and Distribution of Special Object", *Transactions of the International Astronomical Union* vol 6 (Cambridge, 1939), 454-457, 455; Jöran Ramberg, "Astronomiska studier i Sydafrika", *PAT* vol 33 (1952), 122-142, 122f.

One of the three Swedish observatories was rapidly and thoroughly modernised according to the style of astronomy developed in the group of stellar astronomy first formed at Uppsala. The instrumental resources and the staff at the Stockholm/Saltsjöbaden observatory was aligned with the Milky Way astronomy. Lundmark was a bit left out. Saltsjöbaden and Uppsala thus became power-houses in Swedish astronomy.

The Many Cultures of Astronomy

DUNÉR, Hasselberg, Charlier, Lundmark, Lindblad, &c. had different ways of doing astronomy. These differences were highlighted and exaggerated during the process of appointing a professorship. The astronomers during this period seldom discussed the relative merits of their different thoughtstyles. There are traces of their scientific ideals hidden in letters and in the way they pursued astronomy; in fact, astronomers mostly did astronomy, rather than discussed how to do it. But when it came to appoint professorial chairs, they became more open about how to weigh the relative merits of the various astronomy subcultures. The process of appointing a professor was, of course, also an arena of power politics within the learned world. In this, the last, chapter, we will follow the arguments in a number of professorial appointments. We will also look at astronomy-related science taking place between physics and astronomy. Then, some astronomical activities in the field of popularisation are discussed, as well as the place of history of astronomy in the astronomical community. To finish the book, there is a discussion of the international contacts of Swedish astronomy, with particular emphasis on the way Swedish astronomers operated within the International Astronomical Union.

Professorial Appointments and Disappointmens

Professorial chairs in Sweden, in astronomy as well as in other fields, were rather few and far between. The chairholder had much to say about the steering of astronomy at that particular institute. Thus, great weight and power were attached to these chairs; the process of appointing a professor was an important part of the distribution of power within Swedish astronomy. To follow the appointments of stellar astronomers is a way to chart their gradual taking over of Swedish astronomy. The rhetorical level was of course much inflated during several of these appointments, but still the arguments say something about how astronomers perceived their science.

The first professor of astronomy to belong to stellar astronomy and astrophysics was Nils Dunér, who became professor at Uppsala in 1888. Professors Gustaf Lundquist (mechanics) and Hugo Hildebrand Hildebrandsson (meteorology) raised the issue of calling Dunér to the Uppsala chair in astronomy, which became vacant when the classicist Herman Schultz retired in 1888. According to the university statutes it was possible to call a scientist to a chair without the post being open to competition if there was an outstanding candidate. Lundquist and Hildebrandsson claimed that Dunér was such a candidate.

A major part of their arguments hinged on Dunér's work with the newest methods of astronomical and astrophysical observations. They pointed out that the recent use of photography instead of visual observations was causing a revolution in the working methods of astronomy. Dunér had personally been involved in this process, with his work with the Carte du Ciel project. They also discussed the observations of stellar and solar spectra made by Dunér; the latter done with an instrument at the very frontier of science: "a spectroscope of enormous dimensions, perhaps the most powerful that has ever been placed on a refractor". With this instrument Dunér had fully proved Doppler's principle. Together with his long-standing position as teacher in astronomy at the Lund observatory and his ardent work on the Lund meridian programme, Dunér's merits in observational astronomy and international recognition (the Institut de France had awarded him the Lalande prize) made him a suitable candidate.[451]

A statement from the classical astronomers Hugo Gyldén (Stockholm) and Axel Möller (Lund) supported the proposal to call Dunér to the professorship. They did hint that Oskar Backlund (classicist) at the Pulkovo observatory might also be a good candidate. They specifically stated that Backlund's strengths lay in the mathematical and mechanical areas of astronomy. It was, however, not possible to compare Dunér and Backlund, since they had "worked on wholly different subjects"; for Gyldén and

[451] Uppsala university archive, philosophical faculty archive, minutes of the mathematical-scientific section, 19 September 1888 §5, appendix, UUB.

Möller, the difference between the new astronomy and the classical was very real and apparent. In the end, they claimed that Dunér should be called; their mention of the classicist and theorist Backlund might have been a way to pave the ground for him in a future professorship competition.[452]

The mathematical-scientific section voted unanimously to call Dunér to the chair. Before this vote was cast, professor of mathematics Göran Dillner claimed that he wanted a more theoretical orientation at Uppsala, since the observational side of astronomy was so well catered for in Lund. Oskar Backlund was a good theoretician, but since Gyldén and Möller in their statement had settled for Dunér, Dillner in the end accepted the choice of Dunér. He wanted more theoretical work (which in those days meant celestial mechanics) and hoped that an extra professorial chair would be created.

Thus, the process leading up to the first astrophysicist and stellar astronomer holding a chair was not without its twists and turns, even though Dunér in the end was called to the chair, which could happen, according to the university statutes, only when there was an outstanding astronomer suitable for the post.

Observational stellar astronomy and astrophysics became rooted at Uppsala, later on this tradition was carried further when Östen Bergstrand became professor after Dunér 1909. At Lund, the competition after Möller's resignation in 1895 was between three classicists, C.V.L. Charlier, Karl Bohlin and Folke Engström. Charlier got the post, and eventually left celestial mechanics to develop his school of stellar statistics. This school became quite distinct in Swedish astronomy, producing a large number of PhD's. However, to some extent it failed to establish itself on the professorial chairs; in 1927, two chairs became vacant in Swedish astronomy, and in both cases, members of the Charlier school lost the competition process. By 1927, Charlier's school of stellar statistics got competition from a kind of empirical stellar astronomy that had its roots in Dunér's work; in both these professorial appointment processes, Östen Bergstrand and to some extent, Hugo von Zeipel, could act against Charlier's group.

Candidates for the Stockholm chair, vacant when Karl Bohlin resigned in 1927, were Walter Gyllenberg, Algot Jenvall, Bertil Lindblad, Knut

---

[452] Uppsala university archive, philosophical faculty archive, minutes of the mathematical-scientific section, 1 October 1888 §1, appendix, UUB.

Lundmark, and Gunnar Malmquist.[453] In his statement, Östen Bergstrand argued that Charlier's students Gyllenberg and Malmquist had worked in a school that too much emphasised the analysis of material collected by other astronomers. Bergstrand had a very empirical ideal of science, and he deemed the work of the Lund group too theoretical. Such a purely statistical way of doing astronomy could give illusory results.[454] Such criticism also came from Hugo von Zeipel, who claimed that stellar statistics, as it was practised at Lund, had the risk of landing too far away from reality in it's extensive interest in the details of mathematical statistics; in fact, von Zeipel argued, it was as a contribution to the discipline of mathematical statistics that their work was really important. The Lund astronomers were criticised for being too mathematical. Bergstrand and von Zeipel were also critical of the empirical work of Gyllenberg (a large astrometrical programme observed at the meridian circle); he had not utilised the latest technologies.

Bergstrand and von Zeipel both criticised the candidates from Lund for having been too passive in relation to Charlier. They argued that they had few ideas of their own, and had mainly worked on the problems their leader had pointed out. They both preferred Bertil Lindblad for the post. Östen Bergstrand pointed at Lindblad's developments of methods of spectral photometry that made it possible to determine the distances to large numbers of stars using fairly small survey-type instruments. Arguments about it not being possible to compete in observational stellar astronomy from Sweden had been raised, but Bergstrand noted that Lindblad's methods made such work possible. The observational astronomer Bergstrand looked favourably upon a scientific technology and a set of methods that made it possible to pursue observational programmes in

---

453 Minutes of the astronomy election committee, 2 February 1927, Minutes of the sections and committees of the Academy, KVA. The way of appointing a professor at the Royal Academy's observatory at Stockholm differed from the way university professors were appointed. The post was not formally announced to be open to applicants; an electoral committee consisting of the chairman and secretary of the academy as well as professor of mathematics Anders Lindstedt, the astronomer Karl Bohlin, C.V.L. Charlier, Hugo von Zeipel, and Östen Bergstrand, met to draw up a list of suitable candidates, who then sent in their CV:s.

454 Östen Bergstrand's statement, no 475, Minutes of the astronomy election committee, 7 May 1927, Minutes of the sections and committees of the Academy, KVA; Hugo von Zeipel's statement, no 479, Minutes of the astronomy election committee, 7 May 1927, Minutes of the sections and committees of the Academy, KVA.

Sweden, using Swedish-sized budgets, that were of international interest. Since the Stockholm observatory was about to be fully reorganised and adapted to the modern demands of observational science, it was important to discuss the candidates in terms of their abilities as observers, Bergstrand noted. von Zeipel's preference for Lindblad was, instead, focused on his theoretical insights. Lindblad was, in von Zeipel's eyes, the only astronomer on the list who had raised above the purely empirical state. He looked positively on Lindblad's use of James Jeans' stellar dynamics and Schwarzschild's work in the theory of radiation. Lindblad had shown that he was familiar with the new physics: he had lectured on the work of Bohr and Planck, as well as thermodynamics, hydrodynamics, and statistical mechanics, areas that, von Zeipel argued, were becoming more and more important as a theoretical astrophysics was about to emerge. He also noted that Lindblad besides his scientific merits also had skills of organisation: he had built up a group of astronomers around him at Uppsala.

With Lindblad's victory it became clear that Charlier's group in Lund was not perceived by all as the cutting edge of astronomy. Empirical astrophysical and stellar astronomy of a kind that integrated observational work with modern physics was successful; Lindblad's style stemmed from an empirical style that had been successful in Swedish astrophysics – Dunér, Hasselberg, Bergstrand were the most important predecessors – and now had been updated with modern ways of analysing spectral and photographic data. As opposed to both Lundmark and Gyllenberg, Lindblad was interested in the theoretical details of modern stellar astronomy and astrophysics; he could thus appeal both to Bergstrand the empiricist and von Zeipel the theoretist.

The attacks on the Charlier school continued on their home turf, when Charlier resigned from the chair on 1 April 1927. The expert committee consisted of Östen Bergstrand, Karl Bohlin, C.V.L. Charlier, and Hugo von Zeipel; applicants were Gyllenberg, Lindblad, Lundmark, and Malmquist; Lindblad withdrew his application when he got the Stockholm chair.[455] Bergstrand attacked the Lund stellar statistical astron-

---

[455] Apparently Charlier had originally wanted an expert committee consisting of Bergstrand or von Zeipel, himself and Arthur Stanley Eddington. The science section of the philosophical faculty had protested that since there were four professors of astronomy in Sweden, no foreigners should be allowed as experts. Axel Corlin to Knut Lundmark 1 December 1927, Lundmark's collection, LUB. According to Cor-

omy group in his report in a style reminiscent of his statement in the Stockholm process. Some of their results "did not correspond to physical reality, they are only mathematical results [*räkneresultat*] that depend on the scope, selection, and nature of the observational material used as a basis for the statistical treatment."[456] Bergstrand furthermore continued his attacks on Gyllenberg's large catalogue of stellar positions observed with the meridian circle. When observing at the eyepiece, Gyllenberg had used an old method instead of the newer forms of micrometer that some astronomers claimed gave a higher precision. He was also critical of the project as such; large programmes of meridian circle observations "in the classical style" were bound to become obsolete, as astronomers began to deploy photographic methods instead.[457] Bergstrand was critical of Malmquist's dependence on the methods deviced by Charlier, and that he had become trapped in his belief in the range of the statistical methods, and had been isolated from what Bergstrand saw as the primary astronomical work: observations. Such criticism also came from von Zeipel, who claimed that Gyllenberg's and Malmquist's work smacked of sterility and formalism.[458] He wanted Lundmark, even though he gave him some critical remarks; sometimes, Lundmark's style of writing could be rather unclear, he claimed.[459]

Charlier, of course, argued for his pupils and attacked the Uppsala candidate Lundmark. The criticism against Lundmark hinged on his perceived lack of precision. He stated that Lundmark lacked "self criticism and the talent for high-level precision that are typical of Gyllenberg's and Malmquist's work and which are necessary if one wants to cross the boundary between speculation and scientific results."[460] Charlier gave Gyllenberg and Malmquist a split first position, while Bohlin argued that Gunnar Malmquist ought to get the chair.

The appointment process then developed in several stages. The Lund faculty and senate voted in favour of Gyllenberg. Lundmark filed a complaint, where he attacked the Lund group. He argued that Charlier and his group were old-fashioned and did not understand astrophysics and

---

lin, both Sven Wicksell and Östen Bergstrand had found it strange that Bohlin was on the committee.

[456] *Handlingar rörande tillsättandet av lediga professorsämbetet i astronomi vid universitetet i Lund 1927-1928* (Lund, 1928), 32.

[457] *Handlingar rörande tillsättandet*, 24.

[458] *Handlingar rörande tillsättandet*, 123.

[459] *Handlingar rörande tillsättandet*, 40-59.

[460] *Handlingar rörande tillsättandet*, 90.

modern stellar astronomy. The group around Charlier had not paid sufficient interest to the direct astronomical arguments, and it had developed an exaggerated belief in the quality of statistical methods. An abundance of mathematics is not necessary for stellar astronomy. Lundmark argued that stellar astronomy was a young subdiscipline of astronomy that did not necessarily need a high level of mathematical sophistication; common sense might as well be used.[461] Further criticism came in two texts sent by Lundmark to the section, where he attacked the insular character of the Lund school, which only to a small degree had taken part in international co-operation and it had been somewhat isolated. He also portrayed the group as very conformist: after covering the author's name on a Lund publication, it would be hard to recognise who had written the paper.[462] Further arguments ensued, where Gyllenberg, Malmquist, and Lundmark exchanged critical remarks.[463] Eventually, when the government convened Lundmark was pronounced the new professor of astronomy at Lund after the assistant university chancellor Edvard Rodhe had proposed that Lundmark's protest should be accepted. He claimed that Bohlin's expert report should not be reckoned with.[464] In the late stage of the process, intervention by Östen Bergstrand and Elis Strömgren perhaps helped weigh the matter over to Lundmark's side. Some days prior to the

---

[461] Lundmark's complaints, Ministry of education, papers [*ecklesiastikdepartementets konseljakter*] 12 January 1929 no 46, RA. Lundmark also claimed that Bohlin ought not to have been a member of the committee "because of his long-standing illness" and that his statement should not be considered when the decision was taken. He wrote about Bohlin's problems with the almanac and his expert's opinion the previous year in the Stockholm appointment process, when Bohlin had voted in a way that was seen as strange (he had deemed Lindblad unqualified for the position). The almanac problems alluded to by Lundmark was an incident some years before. Karl Bohlin and the Stockholm observatory had as its duty to publish the Swedish almanac; one edition of the almanac in the early 1920's had an obvious error, which rendered Bohlin much public scorn. The almanac and Stockholm appointment process are two parts of Lundmark's complaint that are marked with a pencil, which hints that they were used as arguments when the government discussed the matter.

[462] Knut Lundmark to the mathematical-scientific section of the Lund University faculty of philosophy 26 September and 2 October 1928, Ministry of education, papers [*ecklesiastikdepartementets konseljakter*] 12 January 1929 no 46, RA.

[463] All in ministry of education, papers [*ecklesiastikdepartementets konseljakter*] 12 January 1929 no 46, RA.

[464] "Universitetskansler, ang. återbesättandet af professuren i astronomi vid universitetet i Lund", ministry of education, papers [*ecklesiastikdepartementets konseljakter*] 12 January 1929 no 46, RA.

decision, Bergstrand had been in contact with the minister of culture, Lindskog. Upon hearing of Rodhes proposal that the government should accept Lundmark's protest note, Strömgren wrote Bergstrand that he thought that the decision was the right one. Also, several foreign astronomers he had been in contact with, agreed. He also stated, that "you have the right, if you think it would further the cause, to use this statement as you wish." Upon receiving Strömgren's letter, Bergstrand immediately sent it to the minister of culture.[465]

Gyllenberg's observational work, the programme of astrometry done with the Lund meridian circle, came under fire with the same arguments when he applied for the post at Uppsala, after Östen Bergstrand had resigned in 1938. Gyllenberg's astrometry was, in the words of Östen Bergstrand, one of the board of experts, a "routine work in the classical style". It lacked in precision; meridian circle astronomy had developed since the 1880's, when the first Lund meridian circle programme was observed; Gyllenberg had not used them in his large project. Bergstrand made a distinction between classical astronomy and "observational work of a more modern style", a category in which he included photographic photometry and spectral analysis, two areas where Gyllenberg lacked experience.[466] Gyllenberg seemed isolated from contemporary stellar astronomy with his emphasis on meridian circle astronomy and stellar statistics, instead of the more vital and modern astrophysical approaches.[467] Bergstrand instead ranked Malmquist first. Malmquist had, to begin with, been Charlier's student and had done work in the style of Charlier's stellar statistics, but later changed to work on the magnitudes, colours and spectra of stars, a large observational activity that had occupied Malmquist since the late 1920's; here, Malmquist had showed an independence from his teacher Charlier.[468]

Like Bergstrand, Lindblad noted in his report the high quality of Malmquist's diligent observational work at Saltsjöbaden, one of which was a "spectrophotometric analysis according to the principles developed

---

[465] Elis Strömgren to Östen Bergstrand 7 January 1928 [!] (copy), Östen Bergstrand to Elis Strömgren 8 January 1929, Strömgren papers, LUB.

[466] *Handlingar rörande befordringsmål vid Uppsala universitet: Professuren i astronomi 1938* (Uppsala, 1938), Östen Bergstrand's report, 5f.

[467] *Handlingar rörande befordringsmål vid Uppsala universitet: Professuren i astronomi 1938* (Uppsala, 1938), Östen Bergstrand's report, 17.

[468] *Handlingar rörande befordringsmål vid Uppsala universitet: Professuren i astronomi 1938* (Uppsala, 1938), Östen Bergstrand's report, 12, 17.

at Uppsala and Stockholm."[469] Lindblad praised Malmquist's mixture of theory and observation.

In his report, Hugo von Zeipel begins with an introduction to the ways of doing stellar astronomy. He contrasts two schools, the modern and the statistical. The beginning of the modern school is dated to about 1915-1920, with the spectroscopic methods of measuring stellar distances.[470] Before the birth of modern stellar astronomy, astronomers had to make do with statistical hypotheses and methods. For example, he attacked Gyllenberg's hypothesis that the absolute magnitudes of stars of the same spectral type had a normal distribution around a mean value. "This hypothesis has really no scientific grounding. It seems to have been used by pure laziness, as it makes certain formulæ easier to handle. Later, after the modernisation of stellar photometry, the truth in this hypothesis has come under attack."[471]

With 121 typewritten pages in folio, Lundmark's expert's opinion is by far the most massive written for the Uppsala professorship.[472] Lundmark argued for Gyllenberg. Lundmark pointed out Gyllenberg's broad qualities; he had mainly worked in classical astronomy but had also done some work in astrophysics, thus he was a more versatile and all-round researcher than the competition. Malmquist had lost momentum since coming to Saltsjöbaden from Lund, and Lundmark also accused him of being a less original and independent scientist, since he had adapted so well to the Uppsala-Saltsjöbaden way of doing stellar astronomy after his arrival there.[473]

---

[469] *Handlingar rörande befordringsmål vid Uppsala universitet: Professuren i astronomi 1938* (Uppsala, 1938), Bertil Lindblad's report, 29.

[470] *Handlingar rörande befordringsmål vid Uppsala universitet: Professuren i astronomi 1938* (Uppsala, 1938), Hugo von Zeipel's report, 2-8.

[471] *Handlingar rörande befordringsmål vid Uppsala universitet: Professuren i astronomi 1938* (Uppsala, 1938), Hugo von Zeipel's report, 41.

[472] *Handlingar rörande befordringsmål vid Uppsala universitet: Professuren i astronomi 1938* (Uppsala, 1938), Knut Lundmark's report.

[473] *Handlingar rörande befordringsmål vid Uppsala universitet: Professuren i astronomi 1938* (Uppsala, 1938), Knut Lundmark's report, 118ff. Lundmark had, as noted above in chapter five, argued for Gyllenberg before. In 1930 he tried to get Gyllenberg as the post of observator at Saltsjöbaden. Lundmark and Gyllenberg had competed for the same chair in Lund, and their personalities were quite different. Perhaps he wanted to get rid of an annoying person at his own institute by arguing for Gyllenberg's case at other institutes. Tensions developed between Lundmark and Gyllenberg. For a period, they were not even on speaking terms with each others. Östen

Malmquist had begun as a pupil of Charlier and later realigned his style of astronomy. Malmquist had begun in stellar statistics but when stellar astronomy moved in another direction, Malmquist had been able to update his research methods. This was viewed favourably by Bergstrand, von Zeipel, and Lindblad, who saw to it that Malmquist became the successor of Bergstrand on the Uppsala chair.

Research in Fields Close to Astronomy

Besides the astronomers working in astronomy proper, a number of scientists from other fields studied the cosmos. Several physicists also published work on astronomy and cosmology. Fields like geophysics had traditional ties to astronomy, and in the field of cosmical physics, some physicists tried to deal with geophysical and cosmological phenomena. Some physicists did this by running empirical programmes, others developed large-scale theories to deal with such datasets. One such theory was advanced by Svante Arrhenius.

Svante Arrhenius (1859-1927) was not an astronomer – he was a physicist and chemist – but he nevertheless developed theories that dealt with cosmology. Arrhenius' work on cosmic phenomena was part of a cosmology which, among other things, dealt with the question of life in the universe and the ultimate fate of the cosmos. Arrhenius' cosmology is very broad; phenomena from the earth's level up to the very largest scales of the universe are fitted to this grand picture of the world. The work of Arrhenius can be seen in a context of work by Swedish scientists for studying geophysical phenomena in the Arctic. Several of Arrhenius' papers and theories sprang from an analysis of data collected by such expeditions to the Arctic; his cosmology can be said to be one theoretical outcome of that great empirical project.

Arrhenius made a number of studies on geophysical and astrophysical phenomena together with the meteorologist Nils Ekholm (1848-1923).[474]

Bergstrand to Knut Lundmark 3 February 1935, Lundmark's collection, LUB; Walter Gyllenberg to Knut Lundmark 8 June 1947, Lundmark's collection, LUB.

474 Elisabeth Crawford, *Arrhenius: From Ionic Theory to the Greenhouse Effect* (Canton, MA., 1996); Anders Ångström, "Ekholm, Nils Gustaf", *SBL*; Axel Wallén, "Nils Gustaf Ekholm", *KVAÅ* 1925, 273-285. For Swedish polar science, see Urban Wråkberg, *Vetenskapens vikingatåg: perspektiv på svensk polarforskning 1860-1930* (Stockholm, 1995); Gösta Liljequist, *High Latitudes: A History of Swedish Polar Travels and Research* (Stockholm, 1993).

During the 1890's, Ekholm was assistant amanuens at the central institute of meteorology, run by the Royal Academy of Sciences. The institute started in 1873. It was a co-ordinator of meteorological observations in Sweden, and also collected observations and predictions from other countries, delivered via telegraphic communication. It issued forecasts that were made available on billboards at stations throughout the railroad system.[475]

Their first study dealt with the influence of the moon on the electrical field in the atmosphere. Arrhenius and Ekholm found that the moon's position to a high degree influenced the amount of electricity in the atmosphere. Their hypothesis was that this was because the moon was negatively charged.[476] The next step concerned the moon's influence on the aurora borealis and the number of thunderstorms. Here they also found a variation with the moon's position.[477] In a third study they discussed a statistical signal in the numbers of aurora and thunderstorms of around 26 days; this period they coupled to the rotation time for the sun's higher parts, the faculæ, at the solar equator.[478]

Arrhenius also fitted man in to the picture in 1898. If geophysical phenomena could vary with the solar and lunar positions, what about humans? Are we also under the influence of cosmic phenomena? Arrhenius set out to look for evidence of such a connection.[479]

To start with, Arrhenius studied the statistics for bronchitis, based on medical data from Stockholm, the "Veckoöfversigt öfver Stockholms sanitära statistik"; no correlation with the lunar position was found. A negative result was also the case with the times of 65 000 deaths.[480]

He did find a correlation between the date of birth and the moon. About 25 000 people's times of birth (the population consisted of members of the society Idun, Swedish military officers and members of Ger-

---

[475] Nils Ekholm, "Meteorologiska centralanstalten", *Nordisk familjebok* (Stockholm, 1913).

[476] Nils Ekholm and Svante Arrhenius, "Über den Einflusss des Mondes auf den elektrischen Zustand der Erde. 1-2", *Bihang till k. svenska vetenskapsakademiens handlingar*, vol 19, afd. I no 8 (1894), vol 20 afd. I no 6 (1895).

[477] Nils Ekholm and Svante Arrhenius, "Über den Einfluss des Mondes auf die Polarlichter und Gewitter", *KVAH*, vol 31 no 2 (1898).

[478] Nils Ekholm and Svante Arrhenius, "Über die nahezu 26-tägige Periode der Polarlichter und Gewitter", *KVAH* vol 31 no 3 (1898).

[479] Svante Arrhenius, "Die Einwerkung kosmischer Einflüsse auf die physiologischen Verhältnisse", *Skandinavisches Archiv für Physiologie* vol 8 (1898), 367-426.

[480] Arrhenius, "Einwerkung", 369.

man nobility) had a period of one tropical month. Two other datasets showed variations: data on 12 000 times of menstruation, collected from Stockholm, and the times of epileptical seizures at a hospital near St Petersburg. The periods were of one tropical month, and one close to 26 days.[481] Arrhenius speculated that, since the atmospheric electrical field also varied, the human nervous system might be influenced by the atmospheric electricity. He also pointed to what seemed like similar phenomena among animals; the palolo worm had a reproductive cycle that was governed by the position of the moon.[482]

Solar and geophysical activity varies in tune, and even man might fit into this network. Something must link these phenomena together. Arrhenius constructed a cosmic physics to account for these findings as well as other cosmic phenomena. His cosmic physics rests on a few physical processes that are used to explain physical phenomena on varying scales throughout the cosmos. In 1900 he discussed such a physical process in a paper on the cause of the aurora borealis: radiation pressure, the force that light exerts on a small particle. It became a central factor in his cosmology.[483]

Arrhenius explains the aurora borealis by dust travelling outward in the solar system from the sun. The theory of Maxwell could be used to show that a radiation pressure exists, as well as allow a quantitative treatment of the problem. He calculated that for a particle with the same density as water, the outward radiation pressure will dominate over gravitation if the particle's diameter is less than 1.5 millionths of a meter. This force pointing away from the sun can be shown to exist by analysing the tails of comets. The Russian astrophysicist Bredichin had studied tails of comets, finding repulsive forces from about the same amount as the sun's gravitation up to about 18 times that. Arrhenius calculated that this corresponded to the radiation pressure from the sun working on particles in sizes from 0.1 to 1.25 millionths of a meter.[484]

In Arrhenius' theory, a very fine dust is issued from the sun; it moves outward in the solar system, sailing on radiation pressure from the sun. The dust comes from the protuberances, the violent eruptions on the sun; when the protuberances have reached above the solar surface, the

---

[481] Arrhenius, "Einwerkung", 375.

[482] Arrhenius, "Einwerkung", 414.

[483] Svante Arrhenius, "Über die Ursache der Nordlichter", *ÖKVAF* vol 57 (1900), 545-580; also in *Phys. Zeitschr.* vol 2 (1900), 81-87, 97-105.

[484] Arrhenius, "Die Ursache", 551f.

matter is condensed into tiny particles of dust. Rays of dust thus issue forth from points on the solar surface, emanating from points on the solar surface where the solar activity is strongest, near sunspots and faculæ. This phenomenon is visible as the rays in the corona, the sun's tenuous atmosphere, observed during solar eclipses. Large parts of the streaming particles fall back on the sun, while some envelope the sun, thus forming the solar corona. Large amounts, however, continue outward in the solar system. For an observer during the night time on the earth, this dust will be visible as the Gegenschein-phenomenon, a diffuse spot of light that is vaguely visible on the point of the sky that is opposite to the sun.[485] Arrhenius argued that the solar dust is electrically charged.[486] The sun issues forth negatively charged dust, while the corona around the sun mainly is positively charged. When the charged dust comes close to the earth, discharge occurs: the aurora borealis.

This theory could explain some periodicities in the occurrence of the aurora. There is a periodicity in auroræ that is about 11 years, the periodicity of sunspot activity. Another variation in the statistics of the aurora borealis is during the year, with maxima at the equinoxes and minima during the solstices. Arrhenius explained these with variations in the earth-sun geometry, as the earth travelled around the sun in it's yearly orbit.[487] Another variation at 25.93 days, discovered by Ekholm and Arrhenius in their earlier analysis on the variation of the aurora; this is explained by the solar rotation period for the faculæ.[488] A periodical variation in the number of displays of aurora has a period of the tropical month. Ekholm and Arrhenius had earlier had the idea that this variation was because of the negative electrical charge of the moon. Here Arrhenius has the same explanation.[489]

What happens to this dust that flows out from the sun? Here, Arrhenius draws up a large-scale panorama of connected cosmic phenomena. Some particles fall back on the sun, while many wander out into space. Some fall on other planets. There is a continuous exchange of matter in the universe, because of the amounts of dust that sail on the radiation pressure of the stars, making the physical states of various places in the universe interconnected. Pieces of the dust emanating from the stars

---

[485] Arrhenius, "Die Ursache", 556ff.
[486] Arrhenius, "Die Ursache", 559f.
[487] Arrhenius, "Die Ursache", 566f.
[488] Arrhenius, "Die Ursache", 568f; Ekholm & Arrhenius, "26-tägige Periode", 19.
[489] Arrhenius, "Die Ursache", 569f.

produce small rocks. One sign of this could be the porous nature of some meteorites, noted by Adolf Erik Nordenskiöld. Other parts of the charged dust reach the nebulæ, where electrical discharge makes the nebulæ shine.[490] Arrhenius had moved from the phenomenon of the aurora borealis to the nature of the nebulae.

The paper on the cause of the aurora was recognised internationally. The mathematical aspects of the theory of radiation pressure was dealt with in more detail by Karl Schwarzschild.[491] The astronomer S.A. Mitchell used Arrhenius' theory to explain the shapes of cometary tails. He also refuted a theory by Bredichin who had thought the forms of comet tails were caused by electrical phenomena.[492]

Arrhenius next turned to the solar corona; after all, it was here that the electrically charged dust that travelled throughout the solar system originated. During a visit to the Lick observatory in 1904, on his way home from the large scientific congress in St Louis, he made a study on the solar corona. He analysed the brightness of the solar corona, as measured during solar eclipses. The feeble light from the solar atmosphere was about as bright as the full moon. With this information and his own theory on the nature of the corona he calculated the amount of dust the corona contained.[493]

---

[490] Arrhenius, "Die Ursache", 576ff.

[491] Karl Schwarzschild, "Der Druck des Lichts auf kleine Kugeln und die Arrhenius'sche Theorie der Cometenschweife", *Sitzungsberichte der math.-phys. Classe der k.b. Akademie der Wissenschaften zu München* vol 31 (1901), 293-338.

[492] S.A. Mitchell, "Comet 1903 Borrelly and light-pressure", *ApJ* vol 20 (1904), 63-68.

[493] Svante Arrhenius, "On the Physical Nature of the Solar Corona", *ApJ* vol 20 (1904), 224-231. The Lick observatory was one of the main observatories doing eclipse work in the early century. The observatory sent out expeditions equipped with large cameras to many solar eclipses, gradually building up a library of plates of solar eclipses. Karl Hufbauer, *Exploring the Sun: Solar Science since Galileo* (Baltimore, 1991); John A. Eddy, "The Schaeberle 40-ft eclipse camera of Lick observatory", *JHA* vol 2 (1971), 1-22; Jeffrey Crelinstein, "William Wallace Campbell and the 'Einstein problem': an observational astronomer confronts the theory of relativity", *HSPS* vol 14 (1983), 1-91; Donald E. Osterbrock, John R. Gustafson, W.J. Shiloh Unruh, *Eye on the Sky: Lick Observatory's first Century* (Berkeley, Los Angeles, and London, 1988). During the short visit (he stayed a couple of days) Arrhenius got into further contact with William Wallace Campbell, which resulted in Arrhenius being part of the Lick observatory expedition to study the total solar eclipse of 1905 from Spain. Arrhenius to Campbell 5 August 1904, copy in Arrhenius collection, KVA

Arrhenius went ahead with synthesising his ideas through a series of lectures on cosmical physics at Stockholm University College.[494] One result of these lectures was his publication of a textbook on cosmical physics, published in 1903 as the *Lehrbuch der kosmischen Physik*.[495] The *Lehrbuch* consists of three main parts: astrophysics, geophysics and atmospheric physics. The astrophysical part contain five chapters: the stars, the solar system, the sun, the planets with their moons, cosmogony. The geophysical part dealt with the mass and form of the earth, the oceans, volcanic activity and so on. In the section about the atmosphere, Arrhenius discussed meteorology, the earth's heat balance, atmospheric electricity, the aurora borealis and the geomagnetic field.

The *Lehrbuch* also contains what became one of the key ideas of Arrhenius' cosmology: stability. The exchange of matter between stars and the nebulæ is a stabilising process, balancing the universe. It stores the energy, keeping the heat death in check. The universe is eternal, the heat death has been countered.[496]

The broad subject matter of the *Lehrbuch* illustrates the definition of the subject of cosmical physics. The first use of the term is not easy to pin down. The astronomer Rudolf Wolf defined the birth of the field as 1852 when Sabine, Gautier and Wolf himself found a variation of the geomagnetic field in tune with the number of sunspots.[497] In 1856 the first textbook of the field was published, when Johannes Müller, professor of physics and technology at the university in Freiburg im Breisgau published a *Lehrbuch der kosmischen Physik*.[498]

The subject of cosmic physics was defined as a mix of astrophysical and geophysical disciplines. Johannes Müller argues that the cosmical physics should study natural phenomena on large scales; it should show how the same laws can explain what is going on in the laboratory, as well

---

[494] *Föreläsningar och öfningar vid Stockholms högskola höstterminen 1899* (Stockholm, 1899).

[495] Svante Arrhenius, *Lehrbuch der kosmischen Physik* 2 vols. (Leipzig, 1903).

[496] Arrhenius, *Lehrbuch*, 221-233.

[497] Rudolf Wolf, *Handbuch der Astronomie ihrer Geschichte und Litteratur* (Zürich 1890-1893), 408f.

[498] Johannes Müller, *Lehrbuch der kosmischen Physik* (Müller & Pouillet, *Lehrbuch der Physik und Meteorologie* volume three) (Braunschweig, 1856) + atlas. Perhaps Müller coined the term cosmical physics; that is at least what Günter claims in his review of Arrhenius' *Lehrbuch*, *Dr. A. Petermanns Mittheilungen aus Justus Perthes' geographischer Anstalt* vol 49 (1903), *Geographischer Literatur-Bericht für 1903*, 152f.

as in the nature at large.[499] Müller saw cosmical physics as the meeting of astronomy with meteorology; also other fields could fit in the picture. In the reviews of the literature in *Beiblätter zu den Annalen der Physik und Chemie*, four subjects were put under the heading of cosmical physics: astrophysics, geophysics, meteorology and studies of geomagnetic phenomena, close to the way in which Arrhenius defined the field in the *Lehrbuch*. The 11th edition of the *Encyclopædia Britannica* defined the field thus

> 'Cosmical physics' is a term broadly applied to the totality of those branches of science which treat of cosmical phenomena and their explanation by the laws of physics. It includes terrestrial magnetism, the tides, meteorology as related to cosmical causes, the aurora, meteoric phenomena, and the physical constitution of the heavenly bodies generally. It differs from astrophysics only in dealing principally with phenomena in their wider aspects, and as the products of physical causes, while astrophysics is more concerned with minute details of observation.[500]

Thus, astrophysics was defined as a more empirical line of work than cosmical physics, and the latter as a more wider subject area that also brought terrestrial phenomena into the picture. Cosmical physics belongs to a synthetic tradition of nineteenth century science, discussed by Susan Faye Cannon and other historians. Cannon found a field that strove to integrate astronomy, geophysics, biology, all with "the goal of discovering quantitative mathematical connections and interrelationships – 'laws,' if you prefer, although they may be charts or graphs", Cannon called the field humboldtian science.[501] Arrhenius style was seen by contemporaries to belong to this field.[502]

Arrhenius' style of doing science was, from the perspective of several precision-oriented Swedish physicists, not very popular. Sven Widmalm has discussed the style of physics that prevailed among the Uppsala school of physics with its emphasis on precision measurement. Proponents of this style – for example Bernhard Hasselberg, physicist and astronomer, and the meteorologist Hugo Hildebrandsson – argued that Arrhenius' work was a sloppy compilation of the work of others. Cosmical

---

[499] Müller, *Lehrbuch*, 2.

[500] "Cosmic", *Encyclopædia Britannica*, 11th ed. (New York, 1910).

[501] Susan Faye Cannon, *Science in Culture: The Early Victorian Period* (New York, 1978), chapter three: "Humboldtian Science". The quote is from page 77.

[502] G. Hellman, review of Arrhenius' *Lehrbuch*, *Meteorologische Zeitschrift* vol 200 (1903), 527f.

physics was not the sole domain of Arrhenius and Ekholm, based at the Stockholm University College; the Uppsala physicists also did cosmical physics, but with a more precision-oriented empirical style.[503]

Arrhenius took his theoretical work on cosmos a step further – and became, in the eyes of the precision-oriented physicists, even more speculative – when he began working on the problem of the origin of life, which he soon developed into an interest in the possibility of life in space. He published his ideas first in *Nordisk tidskrift* in 1905.[504]

Life could not, according to Arrhenius, come out of inorganic material.[505] The only possibility left "is the so-called 'panspermia', according to which seeds of life wander through the cosmos; it meets planets and fills their surface with life, as soon as conditions for the existence of life have emerged on them."[506] Arrhenius mentions several predecessors, panspermia was not a new idea; Arrhenius rather added a new mechanism for transportation.[507] Collissions between planets and transport via meteorites were two transport mechanisms discussed earlier. Richter and others had thought the life spores were transported via meteorites; Arrhenius argued that the mechanism would destroy life as the meteorite was heated upon entry in the atmosphere. Thomson had the idea that collision between planets would case fragments with life to spread to other planets; Arrhenius thought the mechanism to be too rare, and also it had the problems with heat.[508] Instead, Arrhenius proposed, life could travel on the radiation pressure. Arrhenius theory of panspermia was connected to his theories in cosmical physics.

Arrhenius noted that Karl Schwarzschild's calculations showed that the bodies that were the most affected by the sun's radiation pressure had a diameter of 0.16μ, which is of the same order of size as many kinds of bacteria in their resting state. Such a bacterium with the density of water could travel from the earth to Mars in 20 days; it would reach Jupiter after 80 days and Neptune after 14 months. The closest star would be reached after 9000 years.[509]

---

[503] Sven Widmalm, *Det öppna laboratoriet*, ch. 4.

[504] Svante Arrhenius, "Lifvets utbredning genom världsrymden", *Nordisk tidskrift* 1905, 189-200.

[505] Arrhenius, "Lifvets utbredning", 192.

[506] Arrhenius, "Lifvets utbredning", 192.

[507] Michael J. Crowe, *The Extraterrestrial Life Debate 1750-1900: The Idea of a Plurality of Worlds from Kant to Lowell* (Cambridge, 1986; 1988), 404f.

[508] Arrhenius, "Lifvets utbredning", 194.

[509] Arrhenius, "Lifvets utbredning", 195.

Emission of bacteria from the earth was not at all impossible: the thing could be accomplished with the use of electrical forces. High up in the atmosphere the air is "filled with negative electrical charge" because of the negatively charged dust arriving from the sun, causing the aurora borealis. A bacterium could therefore easily pick up a negative electrical charge, and the be repelled from the other electrical particles, to start its drifting in the sea of space.[510]

Arrhenius pointed out that it was not impossible for the bacteria to survive the harsh conditions in space. Bacteria would indeed survive ultraviolet radiation and the intense cold; the latter was even a positive thing, since it would make the life processes of the bacterium proceed slower, therefore conserving the life in the bacterium for longer periods of time.[511]

Arrhenius' work on the origin of life on Earth probably was the main reason why his book *Världarnas utveckling*, published in 1906, became such a bestseller. *Världarnas utveckling* is an introduction to his cosmical physics on a popular level. The book follows roughly the outline of the *Lehrbuch der kosmischen Physik* but on a popular level.[512]

With his widely known ideas about life in the universe, Arrhenius was bound to enter the debate on the canals on Mars. The planet Mars is close to the earth every second year, when it passes on the outside of the earth; a phenomenon astronomers call opposition. Because of the eccentric shape of Mars' orbit, some oppositions are closer than others; these passages occur with an interval of about 15 years. During such a close passage in 1877 Mars two moons were discovered, and Giovanni Schiaparelli found the canals of Mars. The discussion of the nature of these structures on the planet then went on for at least thirty years among astronomers, amateur astronomers, and the public.[513] During the 1909 opposition of Mars, Arrhenius entered the debate on the sceptic side.

During the opposition in 1909, critics of the idea that martian canals were signs of intelligent life had become quite vocal, not the least because of the observations by Antoniadi; he seemed to observe how the canals

---

[510] Arrhenius, "Lifvets utbredning", 197.

[511] Arrhenius, "Lifvets utbredning", 196.

[512] Cf. Olov Amelin, "Physics as Ideology: Svante Arrhenius as a Writer of Popular Science", in Svante Lindqvist ed., *Center on the Periphery*, 42-57.

[513] The issue of the Mars canals is dealt with in Crowe, *Extraterrestrial life*, 480-546; Steven J. Dick, *The Biological Universe: The Twentieth-Century Extraterrestrial Life Debate and the Limits of Science* (Cambridge, 1996), 59-105.

dissolved into smaller, natural spots on the surface of the planet. Astrophysical observations were also used to study the red planet. How to make sense of the astrophysicists' observations of Mars' spectrum was still debated. Arrhenius entered the debate – as was his general style – not as an observer but as a theorist working with other scientists' data.

In an article published in *Nordisk tidskrift* Arrhenius discussed the findings of the 1909 opposition.[514] The main issue was how to interpret the observations of the spectrum of Mars. Several astronomers had, also before 1909, observed spectroscopic signs of water in Mars' atmosphere: William Huggins, Jules Janssen, Angelo Secchi, Carl Hermann Vogel, Edward Walter Maunder; Campbell of the Lick observatory had been critical of these observations.[515] The debate hinged, however, on more recent work by the Lowell observatory. There, spectra taken by Vesto Melvin Slipher showed signs of absorption bands of water in the spectrum of Mars. The results from Lowell and Slipher prompted Campbell to make a comeback in the debate, observing the spectrum of Mars from Mount Whitney, 4420 meters above sea-level.

Arrhenius' view was that the spectra of Mars showed the planet to be unfit for life. His analysis of the spectral data showed the temperature to be about 5 degrees C. at the equator at noon at the most – if Frank Very's measurements of spectra were used – and -17 degrees C. at the least. Mars was dead.[516]

The canals were, however, real to Arrhenius; his view was that they were physical structures on the surface of the planet. He pointed to the success at Lowell observatory of photographing some of the canals in 1909. But he also accepted Antoniadi's view: the very straight appearance of the canals was something that the observer added in his mind; actually, the canals are, if we could see them close up, quite irregular. They were in constant change; fault lines were moved because of marsquakes; sand blew over them; sometimes lakes appeared, caused by the temporal outgassing of water vapour from the interior of the planet.[517] An article by Arrhenius in the German journal *Kosmos* prompted the ageing Schiaparelli to write that he agreed with Arrhenius about the geological interpre-

---

[514] Svante Arrhenius, "Nya undersökningar om planeten Mars", *Nordisk tidskrift* 1910, 81-98. The subject of life in space seemed to thrive in semipopular journals rather than in truly scientific journals.

[515] Crowe, *Extraterrestrial Life*, 504; Steven Dick, *Biological Universe*, 88ff.

[516] Arrhenius, "Undersökningar", 84f.

[517] Arrhenius, "Undersökningar", 97f.

tation of the canals (a word he wished to avoid); however, he thought that periodic changes in the appearance of the structure could be caused by biology.[518]

In a book published in 1915, Arrhenius argued that the planet Venus was quite different from Mars. While Mars was seen as a dead planet, Venus was filled with lifeforms. He calculated the mean temperature to be 47 degrees, and thought that there were large swamps. Temperature is fairly constant so life there has not been under much pressure to evolve; most of it would probably be lower lifeforms, such as plants.[519]

Arrhenius thought of the three planets as representing different evolutionary stages. Young Venus was rich in life, but not very evolved; the earth had, of course, advanced life, whereas Mars could not harbour higher life at all. He thought that life on Venus someday will evolve; the high temperature there will become more moderate and life there will be more advanced. In such a future, our planet will become desolate, maybe because the sun's heat output will diminish. Mars might very well have had advanced life before.The earth will someday share Mars' destiny, however after almost immeasurably long periods of time.

Arrhenius continued writing on the idea of extraterrestrial life. In his last paper – dated four months before his death – he argued that the so-called thermophilic bacteria originate on Venus. Such bacteria thrive at temperatures between 40 and 80 degrees C., which, Arrhenius observed, was close to the temperature of Venus. "Here we have an experimental proof for the transfer of small bodies from Venus to the Earth, travelling on the solar radiation pressure."[520] The bacteria in the biologist's laboratory had originated on Venus.

*

It has not been unusual to find astronomers doing meteorological observations, observations of the aurora etc. Both physicists and astronomers have tackled problems like spectroscopy, solar physics or geophysics.[521]

---

[518] Svante Arrhenius, "Neues vom Mars", *Kosmos* vol. 7 (1910), 123-128; "G. Schiaparelli über die Marstheorie von Svante Arrhenius", *Kosmos* vol. 7 (1910), 303.

[519] Svante Arrhenius, *Stjärnornas öden* (Stockholm, 1915), 150f.

[520] Svante Arrhenius, "Die thermophilen Bakterien und der Strahlungsdruch der Sonne", *Zeitschrift für physikalischen Chemie*, Cohen-Festband, 1927, 516-519.

[521] The Uppsala physicist Knut Ångström worked on measuring solar radiation. He developed an instrument for measuring the amount of solar energy reaching the

But work in this trading zone was sometimes contested. When Carl Vilhelm Ludvig Charlier intended to begin work on seismology at the Lund observatory, several professors opposed the idea, claiming that seismology did not belong to astronomy (see chapter three). Axel Corlin was also exposed to the problems that could arise when an astronomer entered the trading zone.

The student Axel Corlin (1896-1965) was catapulted to astronomical stardom (at least in Sweden) when he discovered a nova on 8 June 1918.[522] Corlin phoned Knut Lundmark at Uppsala observatory, who began making observations of the new star. As it turned out, the nova had been observed at about the same time in other parts of the world, but Corlin nevertheless was one of the first to observe this new star and report it to professional astronomers.[523] This made Corlin, who by that time was a student of astronomy at Uppsala quite famous among people of his hometown Gothenburg, where he was recognised on the streets. Corlin was also being recognised by the Swedish Royal Academy of Sciences, who awarded a gold medal to the young discoverer in 1920.[524] He also began writing on astronomical subjects for the daily press and magazines.[525] These writings provided his main source of income for a time.[526]

Corlin eventually found his way to the Lund observatory, where he, after a number of years, began working on cosmic rays, the radiation whose origins still were a mystery. One early hypothesis held by Corlin was that they originated from variable stars of the Mira type.[527] This idea

---

earth, and he was active in the Solar Union, organised by George Ellery Hale. Sven Widmalm, *Det öppna laboratoriet*, ch. 5.

[522] Entry in Wilhelm Norlind & Bror Olsson, *Lunds universitets matrikel 1939* (Lund, 1940); curriculum vitæ in Axel Corlin's letter to Lundmark, no date, Knut Lundmark's papers, LUB.

[523] "Nova Aquilæ 3", *AN* no 4949 vol 207 (1918); Knut Lundmark, "Mitteilung über die Nova Aquilæ 3", *AN* no 4950 vol 207 (1918).

[524] Axel Corlin to Knut Lundmark June 16 1918, Lundmark's papers, LUB.

[525] Corlin discussed such popular writings in several letters to Lundmark during the years 1918-1920 in the Lundmark papers at LUB. Corlin for example wrote about Lundmark's work at Uppsala in the newspapers. The Lundmark-Corlin correspondence, Lundmark's papers, LUB.

[526] Axel Corlin to Östen Bergstrand 12 August 1923, Bergstrand's collection, NC736, UUB.

[527] Axel Corlin to Östen Bergstrand 24 October 1926, Bergstrand's collection NC736, UUB; Axel Corlin to Knut Lundmark 23-24 March 1927, Lundmark's papers, LUB. The cautious Bergstrand warned Corlin about jumping to conclusions

did attract international recognition, but it was also criticised. A study of Corlin's proposal by B.P. Gerasimovic ended in a conclusion that the Mira variables probably were not the source for the radiation.[528] While the early work by Corlin dealt with statistical discussions of measurements made by Werner Kolhörster and other scientists, Corlin soon moved into doing observations himself.[529] Here he worked in an old Swedish tradition in science: that of making use of the country's far northern position.[530] Some of the scientists who early on were active in cosmic ray physics – Kolhörster and V.F. Hess – had suggested the importance of observing cosmic rays in the north, where auroræ and geomagnetic disturbances were common. Corlin made several observation runs of cosmic rays. Some observations were made on steamers running between Sassnitz and Trelleborg, and between Malmö and Luleå. Many measurements were made at a scientific station at Abisko in the northern part of Sweden, run by the Royal Academy of Sciences. Here, he also measured the amount of cosmic rays that penetrated iron ore, when he brought his apparatus down in the mine at Kiirunavaara. He mounted the ionisation cylinder, as well as the auxiliary instruments, lead shields, batteries etc., on a railway trolley used by the miners, to produce an underground cosmic ray observatory on wheels. However, radiation from an unexpected amount of radium in the ore and radioactive air in the mine galleries disturbed measurements.[531]

Corlin had been introduced to the field and picked up the necessary knowledge from scientists working in the field of cosmic ray physics.

---

about the origin of the rays. Axel Corlin to Östen Bergstrand 27 November 1926, Bergstrand's collection, NC736, UUB.

[528] B.P. Gerasimovic, "On Mira Variables and Penetrating Radiation", *Harvard College Observatory Bulletin* no 847 (1927), 1-5.

[529] Axel Corlin, "The variations with sidereal time in the instensity of highly penetrating cosmic radiation", *Arkiv* vol 21 B no 1(1928); Axel Corlin, "Statistische Behandlung der Höhenstrahlungsmessungen", *Zeitschrift für Physik* vol 50 (1928), 808-848. One of Corlin's conclusions was that the variation with sidereal time was real, therefore implying a cosmic origin of the rays. He also argued that the rays consists of two components, one harder that comes from all directions, one softer that comes from discrete sources and that is the source of the diurnal variation observed in the cosmic rays.

[530] Cf Svante Lindqvist, "Ett experiment år 1744 rörande norrskenets natur", in Gunnar Broberg, Gunnar Eriksson, Karin Johannisson eds., *Kunskapens trädgårdar: Om institutioner och institutionaliseringar i vetenskapen och livet* (Stockholm, 1988).

[531] Axel Corlin, *Cosmic Ultra-Radiation in Northern Sweden* (*Annals of the Observatory of Lund*, 4) (Lund, 1934), A33f, A44ff, A97ff.

They were non-astronomers, mainly Kolhörster at Potsdam; he had borrowed some of the apparatus from K. Büttner of Kiel, and E. Steinke of Königsberg had also delivered an apparatus. He discussed measurements with other participants in the field like J. Clay of Amsterdam, V.F. Hess of Innsbruck, E. Regener of Stuttgart and Steinke.[532] In 1930, Corlin had met with Hess, Kolhörster and Steinke in Berlin; they had discussed simultaneous observations of cosmic rays at stations in different parts of the world. They tried to set up an international cooperative effort of observing cosmic ultra-radiation, as Corlin called the phenomenon. Initial observations showed large variations. To find out how much of these differences were being caused by the fact that observers at different sites used equipment with different characteristics, "exactly similar standard apparatuses are necessary". Such were produced by Steinke at the Physical Institute at Königsberg.[533] Copies of Steinke's standard apparatus were produced and distributed to observers around the world, among them Corlin.[534] Corlin had been hooked up to an international network of physicists studying cosmic ray physics. As often is the case in science, the scientists active in such networks try to develop standardised units, detectors etc., to facilitate the operation of the network. The transportation of such material artifacts as the Steinke standard apparatus, methods of measurement etc., made Corlin belong to the developing international network of cosmic ray physics; as it were, this gave him a disciplinary identity.[535] The problem for Corlin was how such an identity could help him advance within Swedish astronomy.

Corlin's PhD thesis consisted of a discussion of the apparatus used, and many measurements. He also discussed the possible origin of the cosmic rays.[536] Cosmic rays most probably did come from the cosmos, as most workers in the field had accepted an extra-terrestrial origin of the rays. The possibility of a solar and a stellar origin for the rays was discussed.

---

[532] Corlin, *Cosmic Ultra-Radiation*, A112.

[533] Axel Corlin, "Measurements of the Cosmic Ultra-Radiation in Northern Sweden", *Lund Observatory Circular* no 6, 1932, 124-132, 125.

[534] Rudolf Steinmaurer, "Erinnerungen an V.F. Hess, den Entdecker der Kosmischen Strahlung, und an die ersten Jahre des Betriebes des Hafelekar-Labors", in Yataro Sekido & Harry Elliot, *Early History of Cosmic Ray Studies* (Dordrecht, 1985), 25f.

[535] Cf. Jan Golinski, *Making Natural Knowledge: Constructivism and the History of Science* (Cambridge, 1998), 76f.

[536] Corlin, *Cosmic Ultra-Radiation*, A107ff.

Even if the cosmic rays did originate from the cosmos, the practice of cosmic ray physics was far away from astronomy. It was more like physics than astronomy. The measurements of radiation of debated origin made it difficult to use cosmic ray physics to learn something about particular astronomical objects, and the apparatus used in the measurements hardly didn't function like the astronomer's traditional instruments. Unlike the application of spectroscopy to astronomical problems, it was uncertain what new facts about celestial objects could be learned from a study of cosmic rays.

Cosmic ray physics was still in its infancy, and Corlin had few if any scientists to collaborate with in Sweden. He could, however, hook himself up to the old tradition of Humboldtian science in the arctic arena. The Abisko station was like the polar expeditions of the nineteenth century, where scientists from different fields worked together.[537]

Corlin's workspace differed somewhat from the one present in astronomical observatories. He used electroscopes, magnets, lead shields, precision-milliampéremeters etc. in an experimental set-up that was more akin to experimental particle physics than astrophysics; he sometimes observed cosmic rays from deep in the underground, where he had to fight radioactive contamination of the measurements. Even if the rays did come from the cosmos and Corlin had various theories about where they originated, some astronomers perceived the field as not belonging to astronomy. This became evident when Lundmark wanted a supporting statement from Bergstrand in getting Corlin a research stipend. Bergstrand could not give such a statement. Bergstrand could not evaluate Corlin's work in "the field of cosmic radiation, that is, physics or geophysics or whatever one would like to call it", since he lacked the proper knowledge in the field; Bertil Lindblad was also uncertain about his ability to issue a statement on the work on cosmic rays.[538] Bergstrand, who had trained both Lundmark and Lindblad as well as several other up-and-

[537] Carl Gustaf Bernhard, *Abisko naturvetenskapliga station* (*Bidrag till Kungl. Vetenskapsakademiens historia* XVII) (Stockholm, 1985).

[538] Östen Bergstrand to Knut Lundmark 11 October 1934, Lundmark's papers, LUB; Bertil Lindblad to Knut Lundmark 15 October 1934, Lundmark's papers, LUB. This was not the first time Bergstrand voiced criticism of Corlin's abilities as an astronomer; he had, criticised him for, as a student, neglecting the basic training of an astronomer. Bergstrand thought, in 1920, that Corlin had a too feeble knowledge of basic mathematics and astronomy, that he had begun independent research to early, without learning the science properly before. Östen Bergstrand to Elis Strömgren 3 October 1920, Strömgren papers, LUB.

coming astronomers and thus played a role in the definition of astronomy, clearly stated that Corlin's work in cosmic ray physics did not belong to astronomy. For Lundmark, cosmic rays belonged to "pure geophysics".[539] Corlin's scientific style did not really fit in to astronomy and astrophysics as it was defined in Sweden in the 1930's. It was seen as geophysics rather than astrophysics. His peripheral style did not make it easy for him to remain in academic astronomy, and he eventually took up work as a teacher in a secondary school.[540]

*

Another example of a scientist working in the trading zone between astronomy and physics was the physicist Bengt Edlén, who managed to solve several problems in astronomical spectroscopy during the 1930's and 1940's. Edlén's work has been studied by Karl Hufbauer, who shows that Edlén made several contribution to astrophysics. Edlén proposed solutions to the problem of spectral lines in the hot stars of the Wolf-Rayet type, as well as in novæ and the solar corona.[541]

The Uppsala physicists Anders Johan Ångström and Robert Thalén worked on spectroscopy, produced spectrum atlases and measured the wavelengths of spectral lines. Both Thalén and Ångström had been astronomers early in their careers but then shifted to physics. Knut Ångström's work on solar radiation was also on the border between astron-

---

[539] Knut Lundmark, *Astronomien i Lund*, 14.

[540] Corlin also worked on the origin of meteorites. He had the idea that the meteorites were made up of smaller particles in space; a reference to Svante Arrhenius' shows where he might have picked up the idea. Corlin argued that particles in interstellar space are electrostatically charged by cosmic ray bursts, a process that leads to the buildup of meteorites from smaller particles. After he had been criticised by Conrad Lönnqvist, Corlin made another attempt at discussing the build-up of meteorites in interstellar space; this time he argued that the particles were charged by ionisation effects of starlight. Axel Corlin, "How larger bodies may be built up out of small particles in interstellar space", *Zeitschrift für Astrophysik*, vol 15 (1938), 239-262; Conrad Lönnqvist, "A Criticism of the Hypothesis of Electrostatic Accumulation of Meteorites", *Zeitschrift für Astrophysik* vol 17 (1939), 344-359; Axel Corlin, "On the Building up of Larger Bodies from small Particles in Interstellar Space", *Zeitschrift für Astrophysik* vol 18 (1939), 1-24.

[541] Karl Hufbauer, "Breakthrough on the Periphery: Bengt Edlén and the Identification of the Coronal Lines, 1939-1945", in Svante Lindqvist ed., *Center on the Periphery*.

omy and physics. Ångström worked on getting measurements of the amount of heat radiation received from the Sun, the solar constant.[542]

Another example is Bernhard Hasselberg, who worked on spectroscopy and photography both as an astronomer and as a physicist. Physicists and astronomers sometimes could gather around a scientific technology or a problem. Scientists from various fields could enter a trading zone where communication between two disciplines or subdisciplines could take place. But these results were not always uncontroversial, as the case of Corlin shows.

Popular Science and History of Astronomy

A massive amount of popular writings on astronomy was published in Sweden during the time period. This literature was written by astronomers ranging from professors like Hugo Gyldén, Östen Bergstrand and Knut Lundmark to more marginal characters like Axel Corlin and Ansgar Roth.[543] Several astronomers also wrote on the history of their subject. The fields of popularisations of astronomy and history of astronomy are subjects worthy of extensive research on its own and is largely left out in this book, but some space will be devoted here to these fields to provide perspective.

The style and contents of popular astronomy were, of course, open to debate. Just like astronomers had their own style of doing science, their

[542] Sven Widmalm, *Det öppna laboratoriet*, chapters 3 and 5.

[543] Thord Silverbark discuss the introduction of relativity in philosophy and in popularisations in the press in *Fysikens filosofi: Diskussioner om Einstein, relativitetsteorin och kvantfysiken i Sverige 1910-1970* (Stockholm & Stehag, 1999). Kjell Jonsson has studied the view of the world presented in the popular writings by astronomers such as Knut Lundmark and Conrad Lönnqvist. Kjell Jonsson, "Physics as Culture: Science and *Weltanschauung* in Inter-War Sweden", in Svante Lindqvist, ed., *Center on the Periphery*; Kjell Jonsson, "Naturvetenskap, världsåskådning och metafysiskt patos i mellankrigstidens Sverige", *Lychnos* 1992, 103-146. Lundmark's very positive attitude towards popularisations is presented by Anita Sundman in *Den befriade himlen: Ett porträtt av Knut Lundmark* (Stockholm, 1988). Lundmark's output of non-academic writing was immense, ranging from large book projects to a stream of news notices, sometimes written to sort out financial trouble. Ansgar Roth was also a prolific writer of popular science in the daily press, who had started out as an astronomer at the Stockholm observatory, but later lost his work in the restructuring of the observatory after Bertil Lindblad became professor (see chapter five above). For more on the popularisation of astronomy, see Johan Kärnfelt, forthcoming.

way of dealing with and writing popular astronomy differed. Some astronomers wrote almost no popular science and concentrated instead on running an observatory and doing science. Others wrote popular science, but of a character that emphasised results that astronomers had reached consensus on, rather than speculative issues like life in the universe. Bertil Lindblad published sparingly in the popular format, mostly in *Populär Astronomisk Tidskrift*, the non-speculative journal of the Swedish Astronomical Society. Östen Bergstrand is also an example of that style. Bergstrand argued that popular astronomy should only deal with those results that had become stable, it should deal with what astronomers had agreed upon. Lundmark, on the other hand, was more eager to publish new findings.[544] Another prolific populariser of astronomy was Nils Nordenmark, who, after a PhD in astronomy, worked in the insurance business. Astronomy was, however, his passion, and he published a translation of Camille Flammarion's *Astronomie populaire*. Nordenmark also worked in the history of astronomy and was instrumental in the founding of the Swedish Astronomical Society.[545] Like Bergstrand, Nordenmark was critical of a too speculative style of astronomy journalism and popularisation. "It is about time that some authority intervenes" he wrote after having read some articles by Ansgar Roth that he called a "jumbled mishmash [that] mislead the public".[546] That authority proved to be Nordenmark himself, who told the editors of the *Svenska dagbladet* that he had the backing of all Swedish astronomers when he protested against Roth's quasi-scientific writings.[547]

In 1919 the Swedish Astronomical Society was formed. It had as an aim to attract amateur astronomers and to further astronomical science. Astronomers initially tried to use the Society as a vehicle for useful scien-

---

[544] For example Bergstrand to Lundmark, 30 August, 6 September, 12 November 1921, Lundmark's collection, LUB.

[545] Anita Sundman, "Nordenmark, Nils Victor Emanuel", *SBL*.

[546] Nordenmark to Östen Bergstrand, 30 June, 1931, Bergstrand's collection, UUB. There was also some friction between the Axel Corlin and Ansgar Roth in the early twenties, both of whom were, at the time, rather peripheral astronomers (who eventually left astronomy, Corlin to work as a secondary school teacher, Roth to pursue science journalism full time). Roth was a rather prolific introducer of Einstein's ideas in Sweden. Tord Silverbark, *Fysikens filosofi.* Corlin claimed that Roth did not understand the Einsteinian theory of relativity enough to write anything more profound on it in the press. Axel Corlin to Östen Bergstrand 12 August 1923, Bergstrand's collection, UUB.

[547] N.V.E. Nordenmark to Lundmark 5 July 1931, Lundmark's collection, LUB.

tific work. The society tried to begin a programme of observing colours and magnitudes of stars, there were plans to set up an observatory with a reflector telescope run by the Society.[548] The Society was also involved when Lundmark applied for money for his nebular classification and cataloguing project in the late 1920's. But the Society did not become a unifying organisation for all astronomically-minded Swedes. There were early on some tensions between the Swedish Astronomical Society with its *Populär Astronomisk Tidskrift* and the *Nordisk Astronomisk Tidsskrift* published in Copenhagen. Walter Gyllenberg did not initially like the plans for a Swedish astronomy magazine and hoped that the "people in Stockholm would grasp how pointless it would be to start a journal of their own".[549] Elis Strömgren, editor of *Nordisk Astronomisk Tidsskrift*, had ideas about merging the two journals, but he also voiced criticism of *Populär astronomisk tidskrift*.[550] The early days of the Swedish Astronomical Society were not at all harmonious. Nordenmark more or less hated Bergstrand, who was kept out of the board for quite some time (Bergstrand was instead active on the board of the *Nordisk Astronomisk Tidsskrift*), and several people had trouble working with boardmembers Vilhelm Carlheim-Gyllensköld and Karl Bohlin.[551]

The Swedish Astronomical Society worked in various ways to try to keep the interest for astronomy growing in Sweden. For a while it ran an observatory at Skansen, the outdoor museum in Stockholm, and it took over a planetarium that had been an attraction at the 1930 Stockholm exhibition. Having such a populariser of astronomy as N.V.E. Nordenmark as one of the founding members made the society into a vehicle for popularisation.

---

[548] The telescope, a reflector of 14 inch aperture, was to be financed by public subscription organised by the Swedish Astronomical Society. The project was aborted. Invitation to subscription, 20 May 1920, signed by the society's chairman Karl Bohlin, Peter Nilson papers.

[549] At least, that was what the *NAT* editor Strömgren claimed Gyllenberg had said. Elis Strömgren to Östen Bergstrand May 6 1920, Bergstrand's collection, NC736, UUB.

[550] Elis Strömgren to Knut Lundmark 15 February 1924, Lundmark papers, LUB; Elis Strömgren to Knut Lundmark 8 December 1921 (copy), Elis Strömgren to Östen Bergstrand 19 October 1920 (copy), Strömgren papers, LUB.

[551] Cf. Östen Bergstrand to Elis Strömgren, 22 October 1920, Strömgren papers, LUB; Elis Strömgren to Knut Lundmark 8 December 1921 (copy), Strömgren papers, LUB.

The Swedish Astronomical Society was seen by some as too much of a society centred on Stockholm and Uppsala. After Lundmark became professor in Lund, he very seldom participated in board meetings with the society. Eventually, he started a society of his own in the southern part of Sweden, the Tycho Brahe Astronomical Society, whose yearbook *Cassiopeia* shows it to have been a society with a different tone.[552] By the end of the 1930's, the unity of the organisations in Swedish astronomy had become a disunity.

*

Astronomers dealt with historical issues in professional contexts as well as when writing popular science. After a period of expansion, diversion and specialisation, a science like astronomy might find a use for history of science as a unifying force to provide perspective in a period of such diversification; since the dawn of astrophysics, astronomy had witnessed a rapid growth and specialisation, and historical heroes could be used as monuments, in front of which astronomers could assemble and manifest a collective identity. Leading characters from the history of astronomy could be seen as providing such unifying concepts around which to assemble a discipline that otherwise consisted of an expanding array of diverging subdisciplinary subcultures. By gathering around the historical heroes, a sense of fragmentation because of the broadening disciplinary mosaic could perhaps be countered. Interest in the history of science can also be conceived in some instances as being a way of advancing the interests of science, for getting cultural and economical resources. An interest in the history of science can also be traced to periods when science is situated in a context where the production of nationalistic heroes is called for.[553]

Svante Arrhenius, the very successful writer of popular cosmology, also wrote a book on the history of astronomy. Following up the success of *Världarnas utveckling*, Arrhenius issued a popular history of astronomy. *Människan inför världsgåtan* was published in 1907. The historiog-

[552] Anita Sundman, *Den befriade himlen*, 95ff.

[553] A sketch of the astronomers' interest in history of science is presented here, without any aspirations of completeness; the subject is suitable for further investigation on its own and in a setting that compares astronomy to other sciences. Gunnar Broberg, "Naturvetenskapsmännen om sin egen historia: halvseklet före Nordström", *Lychnos* 1983, 120-130.

raphy in *Människan inför världsgåtan* is basically evolutionary. Looked upon casually, it might seem that one scientific system replaces another in a revolutionary fashion; this is not, Arrhenius argues, the case. All production of knowledge is linked together and "grows like a strong tree up from a tiny seed".[554] And Arrhenius found forerunners, pioneers in history that had grasped truths earlier than other people. Arrhenius points out that Amenhotep had come close to the truth, when he put the sun in an important place; Zoroaster and the gnostic thinkers had had ideas similar to the modern idea that the universe is infinite and eternal. And even though there were cultures throughout history that had not grasped much, it was important for Arrhenius to identify these pioneers, especially when their ideas were similar to his own. His theories of the exchange of heat between hot stars and cold nebulæ was a truth that the foremost of ancient mythologies had grasped: it was included in the cosmogony of the old Scandinavians.[555] Siding with the Vikings, Arrhenius could claim to have a strong following on his side.

In his view of the sciences' role in history, Arrhenius also included the benefits to society. Those people that understood the beauty of Nature and the scientific study of Nature would never use treason or violence to gain advantages over their neighbours.[556] One theme throughout his history is that of how to judge the merits of philosophers. Arrhenius argues that philosophers' views of nature have never been very good and their cosmologies have been of lesser worth than the scientists'. Philosophers sometimes happened to hit it right with their speculations, but just as often they ended up wrong. The "poems of creation" of the philosophers had no physical grounding and their philosophical concepts had to be "purified" through science.[557] Arrhenius had worked at the Stockholm Högskola before going to the Nobel Institute for Physical Chemistry in 1905. The Högskola was more geared to the demands of the newly industrialised Sweden than the traditional universities at Uppsala and Lund, where philosophy and theology were more important. It was a place

[554] Svante Arrhenius, *Människan inför världsgåtan*, second edition, (Stockholm, 1907), 128.

[555] Arrhenius, *Människan*, 30ff.

[556] Arrhenius, *Människan*, 177.

[557] Arrhenius, *Människan*, 174, 146 ("poems"), 153 ("purify").

were research was focused on science, not the humanities. This is one background for Arrhenius' way of writing history of science.[558]

The period was one of nationalistic celebration that took many forms, one of which was the idolisation of Sweden's great men of science. One of these heroes was Emanuel Swedenborg, the earthly remains of whom were brought to Sweden in 1908 from England, to a resting place nearby other great Swedish thinkers in the cathedral at Uppsala.[559] Part of this large interest for Swedenborg was centred at the Royal Academy of Sciences, where a collection of Swedenborg's scientific texts was published 1907-1911. It was edited by Alfred Stroh, who also in other ways worked on the Swedenborg monument; he was a driving force in moving Swedenborg to a grave in Uppsala. To the edition of Swedenborg's text were added commentaries by leading experts, the commentary on cosmology was written by Arrhenius.[560]

Arrhenius' commentary is focused on Swedenborg's cosmology in relation to later research. Swedenborg is seen as a forerunner of many ideas that were later picked up by cosmologists like Kant and Laplace, G.H. Darwin, Wright, Lambert; he sees Swedenborg's cosmology as a link between ancient philosophers and Descartes and Kant.[561] Thus, Arrhenius gave his contribution to the project of establishing national heroes.

Tycho, the only megastar in the history of astronomy (so far) coming from Scandinavia, was bound to attract the interest of astronomers. His place of birth, Knutstorp, and the island of Ven, where he worked as an astronomer, had both been Danish but later became Swedish as the outcome of war – this contributed to the use of Tycho by both Swedish and Danish astronomers. Astronomers celebrated Tycho's memory by issuing commemorative books, holding ceremonies at the site of Tycho's observatories, arguing for excavations of what remained of the observatories at Ven, making plans for more or less complete restorations of the observatories, and putting up commemorative plaques there. In effect, they wanted to raise a monument over the Great Dane. This multimedia

[558] Sven Widmalm, *Det öppna laboratoriet*, ch. 4; Gunnar Eriksson, *Kartläggarna*, 17.

[559] Gunnar Broberg, "Naturvetenskapsmännen", 121f, 127.

[560] Emanuel Swedenborg, *Opera quaedam aut inedita aut obsoleta de rebus naturalibus*, I-III (Stockholm, 1907-1911). The commentaries were also published separately, for example Svante Arrhenius, *Emanuel Swedenborg as a Cosmologist* (*Emanuel Swedenborg as a Scientist: Miscellaneous Contributions Edited by Alfred H. Stroh*, vol 1, sectio 3) (Stockholm, 1908). On Stroh, see "Stroh, Alfred Henry", *SMoK*.

[561] Arrhenius, *Emanuel Swedenborg*, 66.

monument - built with bronze, bricks and mortar, paper, staged ceremonies, ideas, even minor planets - was a way to make the astronomers visible. It was the social construction of a cultural resource aimed at advancing the astronomers' cause and putting resources in their hands. It was about making astronomy visible in culture. This fits in with what was going on in other sciences at the time.[562]

When the Danish astronomer d'Arrest visited Ven in 1868, not much was left of Uraniborg and Stjärneborg. Later, in the autumn of 1900, the state of affairs was even worse. Charlier visited Ven together with Hans Hildebrand, custodian of the national monuments (*riksantikvarie*). A few depressions in the ground hinted at the places where some of the underground instrument chambers had been: it was almost impossible to see signs of Tycho's activities. The Royal Academy of Sciences in Stockholm assembled a committee to see to it that the ruins were looked after properly. The committee worked out a proposal for a commemorative ceremony on Ven to be held in 1901, as well as how the ruins were to be protected in the best way. The site was to be marked by two commemorative plaques. Uraniborg was to be excavated properly, and the ruins protected; Stjärneborg was also to be excavated, the water was to be drained and a protective building built upon it. Astronomers then spent decades trying to make something of these plans.

An excavation was decided on by Hildebrand, so that in the year of the Tycho jubilee, visitors to Ven could get an impression of the extent of the observatories. The excavation proceeded during the summer of 1901 under the scientific supervision of Charlier; a number of artifacts were discovered and Tycho's subterranean observing chambers once again appeared. The day of the Tycho jubilee arrived. In Lund a bust of Tycho was inaugurated. In Stockholm, the Royal Academy of Sciences held a meeting where Nils Dunér held a speech celebrating Tycho. Bernhard Hasselberg launched his facsimile edition of Tycho's *Astronomia instauratæ mechanica*, complete with a Latin introduction written by Hasselberg. On Ven, the ruins were temporarily exposed and visible, after the jubilee they were filled with sand as a protective measure.[563]

---

[562] Gunnar Broberg, "Naturvetenskapsmännen", 122.

[563] Heinrich Louis d'Arrest, "Die Ruinen von Uranienborg und Stjerneborg im Sommer 1868", *AN* vol 72 (1868) No 1718, 209-224; C.V.L. Charlier, *Utgräfningarna af Tycho Brahes observatorier på ön Hven sommaren 1901* (Lund, 1901); N.V.E. Nordenmark, "Stjärneborgsproblemets historia under femtio år", *PAT* vol 32 (1951), 87-93.

Other ways of celebrating Tycho were more drastic. The Copenhagen astronomer Thiele proposed that Stjärneborg should rise again as a modern observatory under joint Scandinavian command: the directorship would rotate between a Swedish, a Danish, and a Norwegian astronomer. This project did attract the attention of at least Charlier. However, such a move did not get approval at the state custodian of monuments. As the historically interested Carlheim-Gyllensköld was later to pronounce, "such an addition of modern elements to an ancient site must abolutely be rejected as serious vandalism, according to all sound modern principles for the preservation of monuments which demands a piety that leaves the ancient sites untouched."[564]

When the international astronomical organisation the Astronomische Gesellschaft convened in Lund in 1904, quite a sizeable time of the conference was devoted to Tycho and the future of the remains on Ven. On the first day of the conference, the board of the AG put flowers by the bust of Tycho that stood in the observatory park. And during a one and a half hour break during the proceedings, astronomers were told that they might very well visit the museum of history to see the collection of tychoniana.[565] On the second day, Bernhard Hasselberg presented his work of publishing the *Astronomia instauratæ mechanica*. Hasselberg had studied a number of copies, and notes of greetings written in these copies gave a picture of Tycho's contacts. Hasselberg also discussed details like the binding and the prints in the book.[566]

The third day of the congress was devoted entirely to Tycho. At 10 am, the steamboat Hven left Malmö harbour.[567] After a nice trip in the lovely weather and a breakfast onboard, the company of astronomers and their wives were transported in a large number of carriages to the site of Uraniborg and Stjärneborg. Charlier gave a presentation of the remains; he noted that the ruins were in a sordid state because of centuries of neglect. Charlier told the astronomers that he wanted the congress to draft a document that lamented the sad state of the ruins.

[564] N.V.E. Nordenmark, "Kring Tycho Brahe-jubileet 1901", *PAT* vol 33 (1952), 15-22, 17; Charlier, *Utgräfningarne*, 19; Vilhelm Carlheim-Gyllensköld, "Om bevarandet av minnena efter Tycho Brahes observatorier på Ven", *PAT* vol 2 (1921), 31-59, quote: 47.

[565] "Astronomkongressen", *Lunds dagblad* 6 September, 1904.

[566] "Astronomkongressen", *Lunds dagblad* 7 September, 1904.

[567] "Astronomkongressen: utfärden till Hven", *Lunds dagblad* September 8 1904.

After this "came an episode of great dramatic effect", as the anonymous reporter of *Lunds dagblad* put it. Johann Palisa said that they were gathered at one of the "sacred places" of astronomy and that some kind of monument was needed to commemorate the achievements of Tycho. He did not, however, propose a regular monument but the naming of a minor planet in honour of Tycho; he had an unnamed asteroid suitable for the task. A glass of water from a fountain used by Tycho christened the minor planet number 499 to Venusia, after the Latin name Tycho used for his island.

The fourth day of the conference also had Tycho on the schedule, when the problematic issue of the state of affairs of the remnants at Ven were discussed.[568] Professor Wislicenus proposed that the board of the Astronomische Gesellschaft should see to it that a report on the status of the Ven ruins was written by experts in advance of every congress of the society. Thiele felt some criticism in this and wanted to change Wislicenus proposal. Perhaps also Nils Dunér and Bernhard Hasselberg felt this criticism; Dunér presented the steps he and the Academy's committee had taken since 1901. The commemorative ceremony had been held and some excavations had been done with funds from the Royal Academy of Sciences and the Royal Academy of Letters, History and Antiquities (*Vitterhets- historie- och antikvitetsakademien*), and the plaques had been mounted at the site; they were paid by a contribution by the king. Dunér also told the conference that Hildebrand by the summer of 1904 had tried to get resources for draining the ruins and putting a fence around the site, but not for the protective building. Dunér also gave hints of his doubts about Hildebrand's commitment.

Charlier told the congress about the activities made by Lund University, a story that also hinged on criticising Hildebrand who, as state custodian of monuments, could dictate what should be excavated. The Royal Physiographic Society, based in Lund, had sent a proposal to the government to give the ruins an adequate shelter; the government remitted it to the custodian of monuments for an opinion. Hildebrand's answer had been that there were so many sites in Sweden that deserved an excavation that Uraniborg's and Stjärneborg's turn would come after about ten years. Charlier had then visited Hildebrand and told him about the international conference in Lund that was about to take place in two years

[568] The following account is based on articles about the AG congress in *Lunds dagblad*, 6-9 september 1904, and N.V.E. Nordenmark, "Kring Tycho Brahe-jubileet 1901", *PAT* vol 33 (1952), 15-22.

time; then Hildebrand had said that the thing came in another light, and promised some action. Nothing had happened. The congress of the Astronomische Gesellschaft eventually voted for a resolution that complained about the sad state of the remnants of Tycho Brahe's observatories.

The conflict between Swedish astronomers and the state custodian of monuments to make a proper excavation at Ven – and in effect turn the place in to a monument of Tycho – was further complicated by the fact that the Danes also felt that they could contribute to the matter. The conflict had been going for a couple of years between Thiele in Copenhagen and some Swedish actors before it surfaced at the Lund meeting. Thiele's proposal in 1901 to reconstruct Stjärneborg had not been very well received in Sweden. During the 1904 Astronomische Gesellschaft congress, the ruins had been on display, with the sand temporarily removed; when state custodian of national monuments Hildebrand was about to order the sand back in (for the second time in three years) he found that the Dane Thiele had done just that. Thiele did not have sufficient permissions from Swedish authorities to conduct archæological work in Sweden, and the sand cover operation was also badly performed, argued Hildebrand, but the issue also struck a nationalistic chord in him: "We must care for our historical heritage without interference from foreign powers" Hildebrand wrote in a report to the Swedish minister of education soon after the conference, in October 1904.[569]

Tycho was a unifying figure for the astronomers, but he could also generate controversy between astronomers because of nationalistic differences, or between astronomers and archaeologists.[570] Ven was an important place when it came to take visiting foreign astronomers somewhere. In this, the sad state of the ruins on Ven could be a nuisance for the Swedish astronmers. Visits to Ven could also be important symbolic acts, as evident in the rows between Lindblad and Lundmark over the excursion to Ven during the 1938 General Assembly of the International Astronomical Union held at Stockholm (see chapter four), or the 1926 Astronomische Gesellschaft congress in Copenhagen, when Elis Strömgren was afraid that Swedish and Danish astronomers and learned societies would be organising two independent and parallel excursions to Ven, a

---

569 N.V.E. Nordenmark, "Kring Tycho Brahe-jubileet 1901", 17f.

570 Not all astronomers were interested in the Tycho issue. Östen Bergstrand was bored with Tycho, thinking that there had been too many jubilees and celebrations. Östen Bergstrand to Elis Strömgren 15 March 1926, Strömgren papers, LUB.

source of confusion and embarassment he dearly wished to avoid.[571] Swedish and Danish astronomers reached out to the island of Ven, situated halfway between Sweden and Danmark, one side not always knowing what the other side was doing, or agreeing on what was to be done.

*

Knut Lundmark was early a prolific writer of popular science in the daily press. If anything, this level of activity increased during his career as astronomer. As he devoted more and more energy to popularisations of astronomy, he also worked more and more on the history of astronomy. The historical perspective was important to him. In the mid-1930's he started an activity in the history of astronomy that was quite ambitious. The series *Historical Notes and Papers* was started at the Lund observatory.[572] Here, Lundmark's own historical work was published, as well as the work of colleagues at the Lund observatory. Lundmark published, among other subjects, on Aristotle as a practical astronomer (*Historical Notes and Papers* no 1), the cosmology of Demokritos (no 2) Greek cosmology (no 4), and the question of the earth hanging free in space (no 3).

His colleagues also published in the series. Björn Svenonius, a graduate student of Lundmark's, had measured the orientation of Snofru's pyramid while working at the Helwan observatory in Egypt. The result was the publication of "On the Orientation of the Egyptian Pyramids" as *Historical Notes and Papers* no. 5. Per Collinder discussed the large-scale development of astronomy in "Statistical Notes on the Astronomers of Antiquity" in *Historical Notes and Papers* no.6. Collinder was interested in the various influences from Phoenicians, Egyptians and Babylonians on Greek astronomy, and tried to find hints of that spread of ideas through statistics of their geographic distribution. Björn Svenonius discussed the possible astronomical contents of Swedish petroglyphs in no. 7 of the series.

Lundmark, ever the enthusiast for new projects, started a society for research in the history of astronomy in 1934, "Samfundet för astronomisk historieforskning". The membership included Swedish astronomers like Bertil Lindblad, N.V.E. Nordenmark and Åke Wallenquist, as well

[571] Elis Strömgren to Östen Bergstrand 24 June 1926, 17 July 1926, Bergstrand's collection, UUB.

[572] The papers in *Historical Notes and Papers* also was published in the series *Meddelanden från Lunds astronomiska observatorium.*

as foreign astronomers like Harlow Shapley. Swedish humanistic scholars included in the membership were Johan Nordström, Sweden's first professor of history of science and ideas, the historian of antiquity Martin P:son Nilsson, the linguist Sigurd Agrell and Theodor Wåhlin, an architect with a strong interest in astronomy. Foreign historians of science in the society's membership were George Sarton and Ernst Zinner. All these members were not active in society, but their membership still shows the ambitious plans Lundmark had for the society. The society is one example of Lundmark's avid interest in the history of astronomy and his attempts at pursuing the interest.[573]

For Lundmark, history was important and he argued that scientists should take history of science seriously. He was worried about what was later called the problem of the two cultures. His main concern was not with humanists avoiding science, but rather scientists without historical sensibilities and knowledge, particularly the history of their own subject:

> A scientist must be interested in the human condition, particularly history, and most of all the historical development in his own science. It is often believed, among scientists, that history of science is the same as chronicles and curiosities or anecdotes, and such 'one can do once one has been retired.' The scientist that lacks an historical sense is in fact much regrettable [*beklagansvärd*].[574]

*

History of astronomy played quite a substantial role in the work of astronomers of the time. They devoted substantial amounts of their time to these historical enterprises. Some astronomers, like Lundmark, saw knowledge in the history of astronomy as really important for professional astronomers. It was, according to Lundmark, important for the scientist to have some kind of historical perspective on the developments of his own science.

Other astronomers did not venture as far as Lundmark in the direction of the humanities. Still, the history of astronomy did belong to their world. Astronomers and astrophysicists like Hasselberg, Dunér and

---

[573] "Handlingar rörande Samfundet för Astron. historieforskning. 1934-1951", LUB

[574] Knut Lundmark, "Enhet och mångfald i universum", in *Olösta gåtor och aktuella problem: sjutiosex uttalanden i Svenska Dagbladets enquête om naturforskningens nuvarande uppgifter och mål* (Stockholm, 1937), 165-175, 174.

Charlier (they can hardly be seen as being quasi-humanists, like Lundmark) did argue for the importance of the memory of Tycho. The great Dane had a place in the world of astronomers. The system of symbols that they surrounded themselves with, a system of values that said something of the way these astronomers perceived their place in culture, contained the history of astronomy.

International Contacts

An attempt will be made here to chart some of the patterns of international contacts of Swedish astronomers and to map the way these astronomers interacted with international astronomy. In the early part of this time-period, Swedish astronomers worked mainly with European and Russian astronomers. This had been the case earlier, during the eighteenth century, Swedish astronomers often collaborated with French and German astronomers.[575] Astronomers often travelled to European observatories to get an insight in the modern methods of astronomical observation and analysis.

The central observatory of Russia in Pulkovo, outside St Petersburg, was an important destination for Swedish astronomers. Several of them worked there during the nineteenth and early twentieth centuries. During that time, the cultural connections between Sweden and St Petersburg where lively. The Russian metropolis was close to Sweden; many people from all walks of life worked there.[576] For astronomers, the central observatory at Pulkovo provided an opportunity; places at the Swedish observatories where rather few, and by applying for a post at Pulkovo, the young astronomer could remain in astronomy.

Some stayed for shorter periods, while others spent major parts of their careers at the Pulkovo observatory. Some Swedes advanced to high positions, and for a while the observatory was a cosmopolitan place, with German and Swedish being the most important languages spoken.

Georg Lindhagen was the first Swedish astronomer that arrived at the Pulkovo observatory, close to St Petersburg. He came there in 1847, after

---

[575] Participation in geodetical projects with european astronomers is documented in Sven Widmalm, *Mellan kartan och verkligheten*.

[576] Bengt Jangfeldt, *Svenska vägar till S:t Petersburg: kapitel ur historien om svenskarna vid Nevans stränder* (Stockholm, 1998).

a recommendation from his teacher, professor Svanberg at Uppsala.[577] After a few months, he got a temporary post which in 1849 was transformed into a permanent position. He worked in the large geodetic project of measuring a degree of arc pursued by the Russian Academy of Science. It ran over 3 000 km, from the Danube to northern Norway; it was started in 1816, and finished in 1855. The concluding phases entailed cooperation with Swedish astronomers, and Lindhagen was Pulkovo's emissary in the dealings with Swedish geodesists and astronomers. Lindhagen's work at Pulkovo was quite successful, and the future looked rather good for him (especially after he married Olga Anna Wilhelmina Struve, one of Wilhelm Struve's daughters). Still, he decided to move back to Sweden. One reason was that he felt homesick, another that a dispute with Otto Struve, director of the observatory and the son of Wilhelm Struve. Lindhagen arrived in Sweden in the summer of 1856.[578]

The next Swede to arrive did stay. Magnus Nyrén came to Pulkovo in 1868, and worked there until his retirement in 1908.[579] Nyrén worked mainly in astrometry, a part of classical astronomy that Pulkovo specialised in. Eventually he became the vice director in 1890.

Bernhard Hasselberg came to Pulkovo in 1872.[580] Among other things he participated in the work of Pulkovo for the transit of Venus in 1874. His main work at Pulkovo was in the field of astrophysics. Pulkovo started an activity in this new area of astronomy and Hasselberg could work here on the spectroscopy of comets with instruments that were much better than those at any other place. Hasselberg extolled the resources available to him at Pulkovo, compared to what was available in Sweden.[581] He stayed until 1889, when he got a post as physicist at the Royal Academy of Sciences in Stockholm.

---

[577] Olle Franzén, "Lindhagen, Daniel Georg", *SBL*; Bernhard Hasselberg, "Daniel Georg Lindhagen", *Levnadsteckningar* vol 4 (1899-1912).

[578] Sven Widmalm, *Mellan kartan och verkligheten*, 390ff; Alan H. Batten, *Resolute and Undertaking Characters: The Lives of Wilhelm and Otto Struve* (Dordrecht, 1988), 30-43.

[579] Åke Wallenquist, "Nyrén, Magnus", *SMoK*.

[580] Sven Ohlon, "Hasselberg, Clas Bernhard", *SBL*.

[581] When photography and spectroscopy later was introduced in Sweden, Hasselberg came with advice to Nils Dunér on the new ways of doing astronomy. Bernhard Hasselberg to Nils Dunér December 7, 1883, February 1, December 6, 1884, LUB.

Oskar Backlund came to Pulkovo in 1878.[582] In 1895 he became director, and stayed in that position until his death in 1916, by which time he had been ennobled. Backlund worked in classical astronomy. He studied the orbit of period comet Encke by numerical methods, for which he set up a temporary calculating institute at the Academy of Sciences in St Petersburg. This work was in part financed by the wealthy industrialist Emanuel Nobel. In connection with this calculating work several other Swedish astronomers came over: Karl Bohlin, K.G. Olsson and Vilhelm Carlheim-Gyllensköld all spent half a year in St Petersburg calculating for Backlund.

Several non-Russian astronomers worked at the Pulkovo observatory. The first director, Wilhelm Struve, was born in Altona, close to Hamburg. Otto Struve, Wilhelm's son and since 1862 the director of the observatory, kept German as the language of the institute; the Swedes working there used German or Swedish. Friction between slavophiles and westerly-oriented people was mirrored in the activities of the observatory. The Struves' leadership was broken at about 1890; Hermann Struve did not get the post as director that was previously thought to be reserved to him; instead, F.A. Bredichin, who sided with the slavophiles, got the post and had orders to make the institute more Russian.[583] The Swedish group acted diplomatic: after the nationalistic group of the academy had ascertained that he would not favour foreign elements, Backlund could become director in 1895.

The Swedish astronomers at Pulkovo were successful. During the major part of the 1880's there were three Swedish astronomers employed; one of whom later became director, another vice director. The observatory, therefore, has played an important role in the history of Swedish astronomy. The number of positions available for an astronomer in Sweden was rather small. Therefore, working in Pulkova was an alternative way of staying in astronomy.

At Pulkovo, the Swedish astronomers where able to work in a type of astronomical institute that was not possible at home. Pulkovo observatory played an important role in introducing Swedish astronomers to new technologies of observing the sky.

---

[582] Knut Lundmark, "Backlund, Johan Oskar (Oskar Andreevitj)", *SBL*; Magnus Nyrén, "Johan Oskar (Oskar Andreevitsch) Backlund", *Levnadsteckningar* vol 5 (1915-1920).

[583] Batten, *Resolute and Undertaking Characters*, 224; Nyrén, "Backlund", 561f.

The international contacts that had often been geared toward Germany, Russia, France, and England, gradually changed. The patterns of communication shifted towards the west, and American astronomy became more important for Swedish astronomers. In 1912, Charlier dedicated the first volume of his *Studies in Stellar Statistics* to Harvard observatory's Edward Pickering, an acknowledgement of the important role the Harvard data gathering had for Charlier's stellar statistical programme; some years later he successfully pleaded for Pickering's election into the Royal Academy of Sciences.[584] During the early twenties, Knut Lundmark and Bertil Lindblad both spent parts of their early career there, and after they had returned to Sweden, they kept in touch with American astronomers. Lundmark even got an offer from the Lick director Campbell, to stay on at Lick, which he turned down.[585] Later, during the 1930's, astronomers such as Yngve Öhman, Erik Holmberg, and Carl Schalén visited US observatories. They worked with instruments, used plate collections, and developed contacts with American astronomers. This westward shift among astronomers was facilitated by the founding of the Sweden-America Foundation in 1919. The initiative to this foundation came from a number of leading Swedish intellectuals and industrialists who wanted to increase the number of contacts between Sweden and USA.[586] Both Lundmark and Lindblad worked at American observatories with backing from the Foundation.

As is well documented, the first world war caused a rupture in international scientific work.[587] A situation of nationalistic competition but with international collaboration and also cosmopolitan rhetoric gave way to more hostile feelings between scientists from different countries. German astronomers were initially not allowed membership in the International Astronomical Union, an organisation for astronomers that was formed after the war on the initiative of George Ellery Hale. Scientists in

---

[584] C.V.L. Charlier, *Studies in Stellar Statistics 1.Constitution of the Milky Way* (Lund, 1912); Charlier to the Royal Academy of Sciences 9 February 1917, Astronomy department's "koncept, institutionshandlingar 1907-1928", 156, LUB; C. Skottsberg ed., *Kungl. Svenska Vetenskapsakademien: Personförteckningar 1916-1955* (Stockholm, 1957), 36.

[585] Östen Bergstrand to Knut Lundmark 16 October 1922, Lundmark's collection, LUB.

[586] Dag Blanck, *Sverige-Amerika Stiftelsen: De första sjuttio åren 1919-1989* (Stockholm, 1989), 6-15.

[587] Sven Widmalm, "Science and Neutrality: The Nobel Prizes of 1919 and Scientific Internationalism in Sweden", *Minerva* vol 33 (1995), 339-360.

Sweden, neutral during the war and with a long history of cultural contacts with Germany, often felt that they should try to act as a broker between Germany and the Allies.[588] The astronomers also took part in this process. In 1920, Charlier wrote an article in the newspaper *Sydsvenska Dagbladet* where he criticised the new scientific world order and argued that the Germans should be allowed to take part in international scientific collaboration again.[589]

Astronomers could see how their German colleagues suffered personally and institutionally when economic and other problems wreaked havoc of parts of the German scientific world.[590] This problematic position of German astronomy was one reason that contributed to the decline of the number of contacts with German astronomy and the increase of contacts with American astronomers. Another reason was the expansion of American astronomy. Scientific institutes like the Mount Wilson observatory, Lick observatory, Yerkes and Harvard observatories, helped push the status of American astronomy. Americans also had a number of very visible astronomers, like George Ellery Hale, Edwin Hubble, Harlow Shapley, Henry Norris Russell, who also served to attract Europeans.

The westward shift in international communications is also evident in the language chosen by astronomers. Astronomers began publishing more and more in English, rather than German or French, shown here in three graphs:[591]

[588] Sven Widmalm, "Science and Neutrality".

[589] C.V.L. Charlier, "Vetenskapen och kriget", *Sydsvenska Dagbladet Snällposten* 17 October1920.

[590] Axel Corlin heard about the problems experienced by German astronomers when Max Wolf visited Lund in 1923. Axel Corlin to Knut Lundmark 21, 24, 25, 26 February 1923, 13 and 15 March 1923, Lundmark's collection, LUB.

[591] Axel Nelson, *Akademiska afhandlingar vid Sveriges universitet och högskolor läsåren 1890/91-1909/10 jämte förteckning öfver svenskars akademiska afhandlingar vid utländska universitet under samma tid* (Uppsala, 1911) and John Tuneld, *Akademiska avhandlingar vid Sveriges universitet och högskolor läsåren 1910/11-1939/40* (Lund, 1945).

Dissertations in astronomy 1890-1939

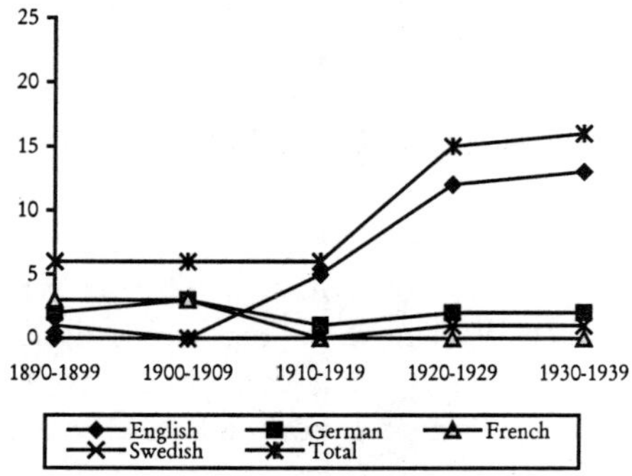

Dissertations in mathematics 1890-1939

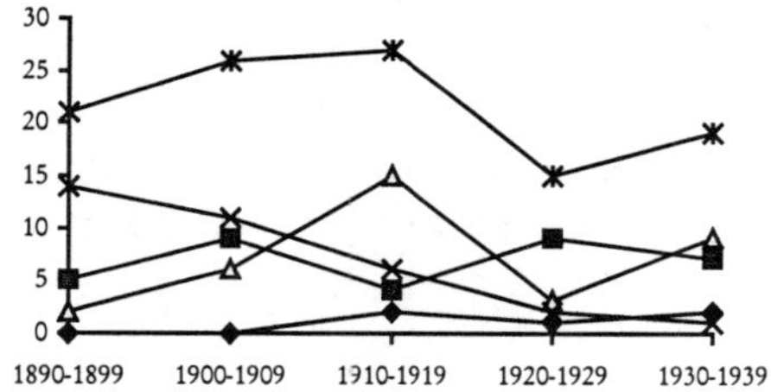

Dissertations in physics 1890-1939

In astronomy, the use of English as the preferred language for the PhD thesis came from 1914. The physics PhD thesis was mainly written in

German up until the end of the 1930's; in mathematics, the languages of choice were German and French. One explanation for the difference between the disciplines could be the different developments in the disciplines. The use of German come as no surprise; Sweden pre-1945 is generally thought of as a society and culture that had many ties to Germany; scientists' travels in Germany and publications using German were very common during the nineteenth century. Astronomy, however, switched to English at least 20 years before physics. A reason for this difference can be the difference in what was perceived as the leading scientific nation, where American astronomy rose earlier than American physics, relatively.[592]

Swedish astronomers worked quite closely with the leading international organizations. Before the war, much astronomical collaboration was done under the auspices of the Astronomische Gesellschaft, based in Germany but in reality an international organization. As noted above, Hugo Gyldén became president of the organization. The International Astronomical Union (IAU) had since its founding in 1919 been an important place were astronomers could interact. It worked as a clearinghouse for new discoveries and it was also an arena where astronomers discussed and decided on matters like standardisation, methods and units.

---

[592] The westward turn was strong, but it was not absolute. During the 1930's, several Lund astronomers worked with telescopes and plate collections in Germany. The ethical implications of sending astronomers to Germany during the second half of the 1930's, and even after the outbreak of the war, seems to have been a non-problem among some astronomers. The Nazi rule is mentioned in passing just a few times in the letters from the young astronomers sent to Germany and Knut Lundmark. Erik Holmberg wrote to Knut Lundmark from Heidelberg telling that the whole staff of the observatory was going to Karlsruhe to hear Hitler's speach and the kick-off of the Nazi election campaign. He also says that he had seen troops march in Heidelberg some days before. The Germans around him were reported to be enthusiastic. Erik Holmberg to Knut Lundmark no date, Lundmark's collection, LUB. Contacts with Nazi Germany did not stop immediately during the war. Knut Lundmark was for example to have participated in an academic meeting in Rostock 24-28 November 1940 (seven and a half months after German troops had occupied Denmark and Norway) together with 27 other Swedish scholars, four of them professors. Something got in the way and Lundmark could not join the meeting. Sverker Oredsson, *Lunds universitet under andra världskriget: Motsättningar, debatter och hjälpinsatser* (*Lunds universitetshistoriska sällskap årsbok 1996*) (Lund, 1996), 101ff. The Lund observatory also organised a trip to Soviet astronomical institutes and observatories in 1936. The visit to the Soviet institutes seems to have lacked major complications, other than the Soviet customs officers confiscated a film from one of the participants. B. Asplind to Knut Lundmark 10 May 1936, Lundmark's collection, LUB.

Every third year, the IAU met and discussed what had been achieved and what were the important issues to be worked on in the near future.[593]

The activities were organised in commissions that dealt with the subdisciplines of astronomy. Prior to the general assemblies every third year, the commissions issued reports of the work done in the various fields. Commissions mapped the various fields of astronomy in an ongoing classification of the science of astronomy. They were not static: commissions could disappear or be created, attract many new people or be looked upon as rather stale, as areas of astronomy grew in size or were deemed uninteresting or obsolete. Commissions in the IAU are examples of how an organisation maps a science that is a conglomerate of various subcultures.

In 1925 Sweden joined the IAU. Swedish astronomy had in the beginning been sceptical of the new organisation, just like several other Scandinavian astronomers. They were critical of the lack of representation by German astronomers in the Union, which they saw as not truly international.

The following table shows the membership of Swedish astronomers in the various committees of the IAU. Bold denotes chairmanship.[594]

Committee	1925	1928	1932	1935	1938
Notations					
Ephemerides					
Bibliography			Lundmark	Lundmark Nordenmark	**Lundmark** Collinder Nordenmark
Telegrams					
Dynamical astronomy		Charlier			
Meridian Astronomy			Gyllenberg	Gyllenberg	Gyllenberg
Instruments					Öhman
Sunspots and Character Figures					
Chromospheric Phenomena					
Solar Radiation and Solar Spectroscopy					
Eclipses					
Wave-lengths					Edlén
Planets and Comets (Physical)					
Lunar Nomenclature					
Longitudes					
Variation of Latitude					
Minor Planets etc (Orbits)				Asplind	Asplind
Meteors					
Carte du Ciel					Lindblad

[593] Adriaan Blaauw, *History of the IAU: The Birth and First Half-Century of the International Astronomical Union* (Dordrecht, 1994).

[594] The information is from *Transactions of the International Astronomical Union* vols 2-6.

Stellar Parallaxes and Proper Motions		Lindblad	Lindblad	Lindblad	Asklöf Lindblad
Stellar Photometry	von Zeipel	Bergstrand Malmquist von Zeipel	Bergstrand Lindblad Malmquist von Zeipel	Bergstrand Lindblad Malmquist von Zeipel	Bergstrand Holm Lindblad Malmquist von Zeipel
Sub Committees on photometry					Lindblad Wallenquist
Double Stars					
Variable Stars	Lundmark				
Nebulae and Clusters of Stars	Lundmark	Lundmark von Zeipel	Lundmark von Zeipel	Lindblad Lundmark Wallenquist	Lindblad Lundmark Wallenquist
Stellar Spectra		Lindblad	Lindblad	Lindblad	Lindblad
Radial Velocities					
Time					
Selected Areas					
Stellar Statistics	**Charlier** Lundmark von Zeipel	Charlier Lindblad Lundmark Malmquist	Charlier Lindblad Lundmark Malmquist	Gyllenberg Lindblad Lundmark Malmquist Schalén	Gyllenberg **Lindblad** Lundmark Malmquist Nordström Schalén Stenquist
Interstellar Matter					Schalén
Solar Parallax					
Stellar Constitution			von Zeipel	von Zeipel	von Zeipel
Spectrophotometry			Öhman	Lindblad Öhman	Lindblad Öhman
Official positions					
Chairman of the union					
Vice chairman of the union				Bergstrand	Bergstrand

The table is an illustration of how few Swedish astronomers worked in classical astronomy. In these, only Walter Gyllenberg (meridian astronomy), B. Asplind (asteroids) and Sten Asklöf were active. Stellar parallaxes dealt not only with trigonometric parallaxes of stars, but spectroscopic parallaxes as well, which explains Lindblad's membership there. The largest numbers of Swedes are in photometric studies of stars and stellar statistics, a name for a commission that during the 1930's dealt with many issues of stellar astronomy that were not solely classical stellar statistics. Stellar photometry and stellar astronomy were related fields; stellar photometry was an essential technology for getting data for models of the stellar system. The concentration of Uppsala and Stockholm astronomers in these commissions hints at a centre of stellar astronomy at Uppsala and Stockholm.

Lindblad and other astronomers doing stellar astronomy were active in commission no 33, stellar statistics. The activities here were not, however, identical to stellar statistics as it was defined by Charlier, Kapteyn and others before the war (and before Shapley's 'large galaxy'). The label stellar statistics was kept for a group of astronomers that worked on the distribution of stars in space but with more modern, astrophysical methods. They tried to measure the distances to stars using astrophysical techniques by establishing luminosity criteria in the spectra of stars. The field was a mixture of various techniques for measuring properties of the stars that could be used for measuring their distribution throughout the galaxy; stellar statistics had been transformed. In 1938 there were 7 Swedish astronomers in commission 33 out of a total of 32. The field was something of a speciality for Swedish astronomy.

Astronomers were proposed as members to the commissions. The commission were arenas where patrons could further the careers of clients, by getting them membership, and proposing that their work was mentioned in commission reports. In his analysis of American astronomy, John Lankford argues that to build a career in the community of American astronomy, both merit and patronage were important. Lankford argues that in some instances patronage was even more important than merit.[595] The commissions of the IAU were arenas where patronage could be distributed on an international scale. Astronomers in the higher spheres of the IAU communicated names of astronomers that they proposed for membership in commissions, sometimes people close to them.[596]

Commissions issued reports that discussed what had happened during three years. These reports show scientists trying to define what they thought was good science. A favourable mention in such a report could be important for astronomers. The chairman of a commission sent out proposals for commission reports in advance of the general assemblies held every third year. In answers to these preliminary reports, leaders of institutions argued for results and practices developed at their institutes to be included in the reports. In their final form, commission reports were printed and presented at the general assembly of the IAU. Commission reports were a regular definition of what counted as good science.

At the IAU general assembly in 1932, Lindblad was elected chairman of the commission for stellar statistics.[597] Lindblad was by then not only the director of a leading Swedish institution but also a certified expert in

---

[595] Lankford, *American Astronomy*, 277.

[596] Lindblad to the general secretary of the IAU F. Stratton 16 December 1932, Lindblad's papers F2:1, KVA.

[597] F Stratton to Lindblad 5 November 1932, Lindblad to Stratton 21 November 1932, Lindblad's papers F2:1, KVA.

his field of work. Lindblad wielded his official power in trying to convince the presidents of other commissions to take notice of the work done at his institute. He wrote to Henry Norris Russell, recommending that he take notice of the work done by Stenquist and himself on the CN-band in stellar spectra, Ramberg's work on luminosity criteria for red dwarves and Öhman's work on stellar spectroscopy. He informed H.H. Plaskett of the work done at Saltsjöbaden in the field of spectral photometry.[598]

Committee work sometimes centred on questions of definition and standardisation. In an evolving field like spectral classification, it was not advisable that observatories used their own schemes for classifying spectra. Some observatories did however develop classification schemes, and the IAU committees had to adopt new recommendations; committees ran an ongoing activity of standardisation as astronomical spectroscopy unfolded.[599]

Lindblad sometimes ventured outside of the area of astronomy when he argued for international recognition. With Russell, he discussed physicist Bengt Edlén's studies of the spectra of Wolf-Rayet stars, one example of the trading zone between astronomy and physics. Astronomers and physicists had a common interest in the interpretation of spectra; Edlén's work was deemed so important for astronomers that he was elected into the IAU commission on wavelengths.[600]

When the observatory in Saltsjöbaden was started, F.H. Seares (photometry commission) and van Rhijn (stellar statistics commission) were informed on the work that was to be performed there. (Lindblad called the programmes to be performed at Saltsjöbaden a continuation of work done at Uppsala and Lund, so as to give an impression of the emergence of a new centre of Swedish astronomy with connections to these observatories rather than the old Stockholm observatory.) Commission presidents were well informed on what was going on, and telling them

---

[598] Lindblad to Russell 17 December 1934, Russell to Lindblad 2 January 1935, Lindblad to H.H. Plaskett 4 January 1935, Lindblad's papers F2:1, KVA.

[599] Lindblad to Walter S. Adams 29 January 1935; "Report of sub-committee on Criteria for Classification of Stellar Spectra", signed by Walter Adams, both in Lindblad's papers, F2:1, KVA. On standardisation in science, see M. Norton Wise ed., *The Values of Precision* (Princeton, 1995).

[600] Lindblad to Russell 14 April, Russell to Lindblad 3 May 1932, Lindblad's papers, F2:1, KVA.

about the new programmes was a way of laying claim to areas of research.[601]

Information flowed in the commissions in a semi-public space. Commissions and their presidents knew much about what was going on, but that information was sometimes of a tentative nature, not yet fit for publication, sometimes more open, as in the published committee reports. Astronomers felt it important that such communication was held in a civilised tone. The author of the committee reports had to weigh in the views of several astronomers, and a diplomatic tone could be applauded.[602] Sometimes authors of committee reports had to be treated kindly, lest they write something critical in their reports.[603]

In a similar way, it was important to find the right tone when writing to a committee president about the results of one's institute. Lindblad could write in a way that was both quite humble and at the same time really brought forward the institute's results. After having filled two typewritten pages with information on progress at Saltsjöbaden, Lindblad concluded a letter to Russell:

> Of course I do not propose that these things should be included at any length or even specifically mentioned in your report, which must of course be very brief compared to the wide field of researches going on. Still I am glad to communicate them, in the hope that they may be of some interest to you in connection with the report.[604]

Correspondents had to strike a balance between politeness and boasting, when it came to discuss the results of one's institute with the commission presidents. Also, when writing as a president about results performed at one's own institute, Lindblad found it important to use a phrasing that was appropriate.[605]

Work in IAU:s commissions was a way of practically handling the ongoing activities in various fields of astronomy. Here, new methods and results were critically discussed and exposed. These commissions were arenas where the world of international astronomy met to deliver infor-

---

601 Lindblad to F.H. Seares 1 February 1932, Lindblad to van Rhijn 11 January 1932, Lindblad's papers, F2:1, KVA.

602 Henry Norris Russell to Lindblad 1 March 1935, Lindblad's papers F2:1, KVA; Jan H. Oort to Lindblad 25 March 1938, Russell to Lindblad 9 March 1938, Lindblad's papers, F2:2, KVA.

603 Sture Holm to Knut Lundmark 12 January 1934, Lundmark's papers, LUB.

604 Lindblad to Russell 17 December 1934, Lindblad's papers, F2:1, KVA.

605 Lindblad to Seares 29 March 1938, Lindblad's papers F2:2, KVA.

mation about recent progress and disseminate information that was vital to the way astronomy developed. The reports on the progress in the field written by the commission presidents were important for how an institute or a research programme was exposed internationally.

Concluding Remarks

Astronomy in Sweden changed markedly during the period studied here. On the whole, astrometry and celestial mechanics of the classical period were left for a focus on stellar astronomy and astrophysics, often of the observational kind. Astronomy also changed in practice, and photography and spectroscopy became two important new technologies. Theoretical astronomy had been fairly strong among the classicists, who made theoretical studies of the orbits in the solar system. The new astronomy was mostly observational.

It is possible to discern an emphasis on instrumentation and methods of observation among the stellar astronomers and the astrophysicists. They valued additions to the arsenal of ways to observe the sky, and they valued astronomers who contributed to the development of new instruments. For them, the merits of various ways of observing the sky were something important.

The introduction of photography entailed a discussion of this new technology; positive thoughts about the technology pointed at the merits of a mechanised imaging technology, whereas astronomers like Dunér and Charlier grappled with how to extract information from the photographic plates. Dunér sometimes voiced concern about the use of photography, and he was critical of the international efforts to produce a photographic map of the sky, since the plates were not readily possible to compare with the visual methods. Photographic plates recorded faint blue stars, whereas much brighter red and yellow stars showed up faint on the plate, or not at all.

Photography in astronomy was not only used for producing 'pictures'. Astronomers often measured plates to get quantitative data. By measuring precisely the positions of stars on the plates, they could attach the new photographic astronomy to an ideal of measurement and precision that had been prevalent in astronomy prior to that.

In spectroscopy, astronomers worked both quantitatively and qualitatively. Dunér's classification of red stars hinged on his classification of the appearance of stellar stars. However, there was also a precision-oriented quantitative approach to spectroscopy, exemplified by Dunér's measurement of the solar rotation from a precise measurement of the Doppler effect on spectral lines. Later on, spectral classification was not done by looking at a photographic plate to estimate the spectral class or luminosity effects in the spectrum. Rather, astronomers measured the photographic plates with photometer and analysed these measurements, often presented as diagrams of the intensity across the spectrum. These tracings were a step towards a more quantitative spectral analysis.

Astronomy during Charlier's era was characterised by an emphasis on the methods of mathematical statistics, and a large effort to make the handling of empirical data through the Lund observatory as efficient as possible. Charlier's work consisted of a drawing together of heterogeneous elements, that met at the Lund *räknekammare*, the laboratory space that, for Charlier and many of his colleagues and students, became the place of place of practice, instead of the turret with its dome and telescope. This office space can be seen as consisting of elements with links to a multitude of places in early twentieth century science and culture: a number of women assistants and students who worked on routine calculations; mechanical calculators developed in a context of mechanisation of the white-collar workplaces; precise and standardised time-signals distributed through radio telegraphy from central stations in Europe (and redistributed on a public clock at the gate of the observatory); astronomical data from observatories transported to Lund on paper or photographic plates.

Astronomers made much use of such elements that perhaps originated "from cultures far beyond the shores of a master equation or an ontological hypothesis."[606] In such a picture it is hard to uphold a strict limit between science and 'context'. Dunér and the photographic astronomers participated in the photographic societies where the photographic technology and its use for science, medicine, art, and just for plain fun was discussed. Photography came to astronomy from the outside, but when it was imported into the science it had to be properly studied and understood; therefore, astronomers like Dunér became experts on photographic processes and thus wrote articles about the relative merits of developers that were of interest to others than astronomers. The statistical technologies of Charlier were developed to solve problems in stellar statistics, but they also resulted in a competence that could show up in other

[606] Peter Galison, *Image and Logic*, 46-63, quote: 52.

areas of society; there was a flow of technologies in and out of astronomy.

As Östen Bergstrand began doing photographic astronomy of a more astrophysical kind around about 1910, he soon trained Bertil Lindblad and Knut Lundmark in this way of doing astronomy. Eventually, something of a school formed at Uppsala around Östen Bergstrand and Bertil Lindblad; a group that observed in photographic photometry and used the methods developed by Lindblad for determining stellar luminosities from small-dispersion spectra. This technology made it possible to estimate the distance to a star by looking for certain spectral features. This group eventually grew stronger, and later in the twenties challenged the statistical school at Lund. Charlier's pupils did not initially have any success in the two professorial vacancies that opened in 1927: Lindblad and Lundmark won these. Gradually, the group around Lindblad at Stockholm/Saltsjöbaden came to have a strong position in Swedish astronomy. This group had many members of the IAU and had the opportunity of a comparatively well-funded activity when the Saltsjöbaden observatory was constructed. It could also concentrate on research, in contrast with university observatories where teaching duties consumed time. This observatory's instrumentation was geared to the empirical programmes run by Lindblad's group: getting spectrophotometric data for large numbers of stars.

Astronomers belonging to the new stellar astronomy, centred at Uppsala and Saltsjöbaden, were critical of the stellar statisticians, who they thought represented a not so modern way of doing astronomy. These astronomers argued that the stellar statistical school had been too removed from the actual observational practice, and that the development of statistical methods was their main interest, not the acquisition of new facts about the stellar system.

Lundmark's work on classifying nebulæ and putting such data together in a general catalogue of nebular data was becoming increasingly outdated after his appointment as professor at Lund in 1929. Lundmark seemed to lose momentum and spent a lot of time on popularisations of astronomy; for him, astronomy was a science that ought to provide a perspective on nature and culture. It was one source for *Weltanschauung*; this is one reason for his increased activities in popular astronomy and history of astronomy.

The empirical style of the Uppsala-Stockholm school of stellar astronomy was not akin to a simple fact-collecting astronomy. The astronomers in this thought collective developed a style where theory and observations were in symbiosis. Schalén and Lindblad were only two of these who pursued theoretially informed observational work or pure theory. Astronomy had developed to a point where observational stellar astronomy and astrophysics were the leading way of pursuing the science. Observations were performed at a leading observatory at Saltsjöbaden and on two out of three chairs sat professors that worked in that tradition. Observational stellar astronomy, influenced by astrophysics, had become the dominating style of astronomy in Sweden.

Abbreviations

AN, *Astronomische Nachrichten*
ApJ, *Astrophysical Journal*
Arkiv, *Arkiv för matematik, astronomi, fysik*
AT, *Astronomisk Tidsskrift*
DSB, *Dictionary of Scientific Biography*
GHA, *General History of Astronomy*
HSPS, *Historical Studies in the Physical and Biological Sciences*
JHA, *Journal for the History of Astronomy*
KVAH, *K. Svenska Vetenskapsakademiens Handlingar*
KVAÅ: *K. Vetenskapsakademiens årsbok*
Levnadsteckningar, *Levnadsteckningar över Kungl. Svenska Vetenskapsakademiens efter år 1854 aflidna ledamöter*
Lund Medd., *Meddelanden från Lunds astronomiska observatorium*
LUÅ, *Lunds Universitets Årsberättelse*
MNRAS, *Monthly Notices of the Royal Astronomical Society*
NAT, *Nordisk Astronomisk Tidsskrift*
PASP, *Publications of the Astronomical Society of the Pacific*
PAT, *Populär Astronomisk Tidskrift*
SBL, *Svensk Biografiskt Lexikon*
SMoK, *Svenska män och kvinnor*
SOU, *Statens Offentliga Utredningar*
Stock. Ann, *Stockholms observatoriums annaler*
UUÅ: *Kongl. Universitetet i Upsala Redogörelse*
ÖKVAF: *Öfversigt af K. Vetenskapsakademiens Förhandlingar*

BIBLIOGRAPHY

Unpublished

Center for the History of Science, Royal Academy of Sciences, Stockholm (KVA)

Svante Arrhenius papers

Bertil Lindblad papers

Minutes [*protokoll*] of the Royal Academy of Sciences

Minutes appendices [*protokollsbilagor*] of the Royal Academy of Sciences

Minutes of the sections and committees of the Academy [*klass- och kommitteprotokoll*]

Lund university library, Lund (LUB)

C.V.L. Charlier papers

Nils Dunér letters (film copy; collection is at UUB)

Knut Lundmark papers

Carl Schalén papers

Elis Strömgren papers

"Handlingar rörande Samfundet för Astron. historieforskning. 1934-1951"

"Koncept, institutionshandlingar. 1907-1930. [Astronomiska institutionen]" 2 volumes, 1907-1928, 1928-1930.

Nils Dunér, "Genommönstring af nora himlen med afseende på Töckenfläckar", Astronomiska observatoriets journaler och anteckningsböcker

"Dagbok för Lunds Observatorium 1868 Nov. 6 - 1875 Aug. 22"

Lund University archive

minutes of the higher university senate [*större konsistoriets protokoll*] philosophical faculty archive, minutes of the mathematical-scientific section

Lund observatory library

C.V.L. Charlier, "Utredning angående pensionering av änkor och barn efter ämbets- och tjänstemän vid Lunds universitet", typewritten MS dated 21 September 1923

Text by Knut Lundmark dated July 7 1942, bound together with a copy of C.V.L. Charlier, *A Plan for an Institute for Theoretical Astronomical Research* (Lund, 1913)

National Archives, Stockholm (RA)

Ministry of education and ecclesiastical affairs council files [*ecklesiastikdepartementets konseljakter*]

Uppsala University Library, Uppsala (UUB)

Östen Bergstrand papers

Bernhard Hasselberg papers

Arvid Odencrantz papers

Uppsala University archive, philosophical faculty archive:

minutes of the mathematical-scientific section

Journals [*dagböcker*] of the mathematical and scientific section, archive of the faculty of philosophy, 1930, FIb:18.

Peter Nilson family

Peter Nilson, "Knut Lundmark and 'Lund General Catalogue of Nebulæ'", unpublished ms

Proposal for a reflector, Karl Bohlin, 20 May 1920

Knut Lundmark, "Till Svenska Astronomiska Sällskapet", 28 December 1927

"Avskrift av Kungl. Maj:ts nådiga brev till statskontoret", copy to the Swedish Astronomical Association 16 May 1929

Private communication

Interview with Nils Hansson, 1993

Interview with Bo Nilsson, August 1997

The Internet
"History of the Johns Hopkins Department of Biostatistics", <http://www.sph.jhu.edu/biostats/history.html> 17 October 1996

Manuscripts
Klaus Staubermann, *Controlling Vision: The Photometry of Karl Friedrich Zöllner* (PhD thesis, Darwin College, Cambridge, 1998)
Sven Widmalm, *Det öppna laboratoriet: Fysikhistoriska fallstudier, 1853-1910* (forthcoming)

Published

1926 års arbetslöshetssakkunniga. Betänkande och förslag ang. arbetslöshetsförsäkring, arbetsförmedling och reservarbeten (Stockholm, 1928) (SOU 1928:9)
1928 års pensionsförsäkringskommitté. Betänkande med förslag rör. revision av den allmänna pensionsförsäkringen (Stockholm, 1934) (SOU 1934:18)
Adams, Walter S. & Kohlschütter, Arnold, "Some Spectral Criteria for the Determination of Absolute Stellar Magnitudes", *ApJ* vol. 40 (1914), 385-398
Althin, Torsten, *Telelandet Sverige* (Stockholm, 1953)
Amelin, Olov, "Physics as Ideology: Svante Arrhenius as a Writer of Popular Science", in Svante Lindqvist ed., *Center on the Periphery*, 42-57
d'Arrest, Heinrich Louis, "Die Ruinen von Uranienborg und Stjerneborg im Sommer 1868", *AN* vol 72 (1868) No 1718, 209-224
Arrhenius, Svante, "Die Einwerkung kosmicher Einflüsse auf die physiologischen Verhältnisse", *Skandinavisches Archiv für Physiologie* vol 8 (1898)
–"Über die Ursache der Nordlichter", *ÖKVAF* vol 57 (1900), 545-580
–"On the Physical Nature of the Solar Corona", *ApJ* vol 20 (1904), 224-231
–*Lehrbuch der kosmichen Physik* 2 vols (Leipzig, 1903)
–"Lifvets utbredning genom världsrymden", *Nordisk tidskrift* 1905, 189-200
–*Människan inför världsgåtan*, second edition, (Stockholm, 1907)
–*Emanuel Swedenborg as a Cosmologist* (*Emanuel Swedenborg as a Scientist: Miscellaneous Contributions Edited by Alfred H. Stroh*, vol 1, sectio 3) (Stockholm, 1908)

–"Nya undersökningar om planeten Mars", *Nordisk tidskrift* 1910, 81-98
–"Neues vom Mars", *Kosmos* vol 7 (1910), 123-128
–*Stjärnornas öden* (Stockholm, 1915)
–"Die thermophilen Bakterien und der Strahlungsdruch der Sonne", *Zeitschrift für physikalischen Chemie*, Cohen-Festband, 1927, 516-519
Astronomiprofessuren i Uppsala 1938. II (Lund, 1938)
"Astronomkongressen", *Lunds dagblad* 6 September, 1904
"Astronomkongressen", *Lunds dagblad* 7 September, 1904
"Astronomkongressen: utfärden till Hven", *Lunds dagblad* 8 September 1904
Avalder, Hugo, "Arbetets vetenskap", *Nordisk tidskrift i organisation* vol 1 (1919), no 9, 118
Batten, Alan H., *Resolute and Undertaking Characters: The Lives of Wilhelm and Otto Struve* (Dordrecht, 1988)
Baum, Richard & Sheehan, William, *In Search of Planet Vulcan: The Ghost in Newton's Clockwork Universe* (New York & London, 1997)
Beckman, Jenny, "Vetenskap och konst: Illustratörsfamiljen Ekblom vid Naturhistoriska riksmuseet", in Marika Hedin & Ulf Larsson eds., *Teknikens landskap: En teknikhistorisk antologi tillägnad Svante Lindqvist* (Stockholm, 1998)
Berendzen, R. and Hart, R., "Adriaan van Maanen's Influence on the Island Universe Theory", *JHA* vol 54 (1973), 46-56, 73-98
Bergstrand, Östen "Nils Christofer Dunér", *SBL*
–*Études sur la distribution de la lumière dans la couronne solaire* (L'éclipse totale de soleil des 20-21 août 1914, deuxième partie, no 2) (Stockholm, 1919)
–"Världsalltets utsträckning", *Nordisk Astronomisk Tidsskrift* new series vol 1 (1920), 33-40
–*Astronomi* (Stockholm, 1925)
–"Nya vidder för celest forskning", *Olösta gåtor och aktuella problem: sjuttiosex uttalanden i Svenska Dagbladets enquête om naturforskningens nuvarande uppgifter och mål* (Stockholm, 1937), 27-30
Bergstrand, Östen & Rosenberg, H., "Vorschlag zur Aufstellung einer Normalsequens zur Bestimmung effektiver Wellenlängen", *AN* vol 215 (1922), 447-452
Bernhard, Carl Gustaf, *Abisko naturvetenskapliga station (Bidrag till Kungl. Vetenskapsakademiens historia* XVII) (Stockholm, 1985)
Betänkande afgifvet den 1 oktober 1898 af komitén för uppgörande af förslag till ny taxa för personbefordringen å statens jernvägar (Stockholm, 1898)

Betänkande i sexualfrågan avg. av befolkningskommissionen (Stockholm, 1936) (SOU: 1936:59)
Betänkande och förslag ang. offentlig arbetsförmedling och statsbidrag till arbetslöshetskassor (Stockholm, 1923) (SOU 1922:59)
Blaauw, Adriaan, *History of the IAU: The Birth and First Half-Century of the International Astronomical Union* (Dordrecht, 1994)
Blanck, Dag, *Sverige-Amerika Stiftelsen: De första sjuttio åren 1919-1989* (Stockholm, 1989)
Block, Henrik, "Sur la stabilité dúne petite planète aux points triangulaires de Mercure", *Arkiv* vol 10 no 4 (1914)
Blomqvist, Göran, *Elfenbenstorn eller statsskepp? Stat, universitet och akademisk frihet i vardag och vision från Agardh till Schück* (*Bibliotheca historica lundensis*, 71) (Lund, 1992)
Bohlin, Karl, [Report of the Stockholm observatory eclipse expedition], *KVAÅ* 1915, 95-97
Bok, Bart J., *The Distribution of the Stars in Space* (Chicago, 1937)
Bok, Bart J. & Bok, Priscilla F., *The Milky Way* (Philadephia 1941; 1945)
Broberg, Gunnar, "Naturvetenskapsmännen om sin egen historia: halvseklet före Nordström", *Lychnos* 1983, 120-130
–*Statlig rasforskning: En historik över rasbiologiska institutet* (*Ugglan: Lund Studies in the History of Science and Ideas*, 4) (Lund, 1995)
Broberg, Gunnar, Eriksson, Gunnar, Johannisson, Karin, eds., *Kunskapens trädgårdar: Om institutioner och institutionaliseringar i vetenskapen och livet* (Stockholm, 1988)
Broberg, Gunnar, Wikander, Ulla, Åmark, Klas, eds., *Svensk historia underifrån: Bryta, bygga, bo* (Stockholm, 1994)
Broberg, Gunnar, Roll-Hansen, Nils, eds., *Eugenics and the Welfare State: Sterilization Policy in Denmark, Sweden, Norway, and Finland* (East Lansing, 1996)
Buchwald, Jed Z., ed., *Scientific Practice: Theories and Stories of Doing Physics* (Chicago & London, 1995)
Byström, Ilse-Beth, "Stockholms observatorium i Saltsjöbaden: Ett byggnadsverk av Axel Anderberg", *AT* vol 15, 152-163
Cannon, Susan Faye, *Science in Culture: The Early Victorian Period* (New York, 1978)
Carlheim-Gyllensköld, Vilhelm, "Om bevarandet av minnena efter Tycho Brahes observatorier på Ven", *PAT* vol 2 (1921), 31-59
Carlsson, Anders & Holmberg, Gustav, "Vilhelm Carlheim-Gyllensköld på Stockholms observatorium", *Lychnos* 1995, 179-189

Charlier, C.V.L., *Astrophotometrische Studien* (Bihang till k. svenska vetenskapsakademiens handlingar vol 14 part I no 2) (Stockholm, 1888)
–*Über die Anwendung der Sternphotographie zu Helligkeitsmessungen der Sterne* (*Publ. der Astron. Geselleschaft*, nr 19) (Leipzig, 1889)
–"Om stjernfotografi", *Fotografisk tidskrift* vol 2 (1889), 251-254
–*Kan religion finnas utan gudomlig uppenbarelse?* (Uppsala, 1895)
–"Ist die Welt endlich oder unendlich in Raum und Zeit?", *Archiv für systematische Philosophie* vol 2 (1896), 477-494
–*Utgräfningarna af Tycho Brahes observatorier på ön Hven sommaren 1901* (Lund, 1901)
–"Über das Fehlergesetz", *Arkiv*, vol 2 (1905), no 8
–*Sannolikhetsberäkningar för assuransföreningen 401 Landskrona* (Landskrona, 1907)
–"Wie eine unendliche Welt aufgebaut sein kann", *Arkiv för matematik, astronomy och fysik* vol 4 no 24 (1908)
–*Grunddragen af den matematiska statistiken* (Lund, 1910)
–*Studies in Stellar Statistics I. Constitution of the Milky Way* (Lund, 1912) (*Lunds universitets årsskrift* n.s. afd. 2 vol 8 no 2)
–"Das Bodesche Gesetz und die sogenannten intramerkuriellen Planeten", *AN* no 4623 vol 193 (1912), 260-272
–*Studies in Stellar Statistics II. The Motion of the Stars* (Lund, 1912-1913) (*Lunds universitets årsskrift* n.f. afd. 2 vol 8 no 4 – *K. Fysiografiska Sällskapets Handlingar* n.f. vol 23 no 4).
–"A Statistical Description of Trientalis Europæa", *Arkiv för botanik* vol 12 (1913) no 14, 2.
–*A Plan for an Institute for Theoretical Astronomical Research* (Lund, 1913)
–"Preliminary Statistics of Nebulæ and Clusters", *Arkiv* vol 9 (1913) no 15
–"Die Abstände der Sterne vom Spektral-Typus B", *AN*, vol 201 no 4801 (1915), 9-12
–*Studies in Stellar Statistics III: The Distances and the Distribution of the Stars of the Spectral Type B* (Uppsala, 1916) (Nova Acta Regiæ Societatis Scientiarum Upsaliensis ser 4 vol 4 no 7 – Meddelande från Lunds Astronomiska Observatorium ser 2 no 14).
–*Studies in Stellar Statistics IV: Stellar Clusters and Related Celestial Phenomena* (Lund, 1918)
–*Vorlesungen über die Grundzüge der mathematischen Statistik* (Lund, 1920)

–"Vetenskapen och kriget", *Sydsvenska Dagbladet Snällposten* 17 October 1920
–*Introduction to Stellar Statistics* (Lund, 1921)
–"Huru en oändlig värld kan vara uppbyggd", *PAT* 1922, 1-8
–"How an Infinite World May be Built Up", *Arkiv* vol 16 (1922) no 22
–*Utredning angående inrättandet av ett svenskt statslotteri* (Malmö, 1922) (SOU 1922:24)
–*Studies in Stellar Statistics V. On the Galaxy of the B-stars* (*Nova acta regiæ societatis scientiarum upsaliensis* vol. extra ordinem editum 1927) (Uppsala, 1926)
–*Application de la théorie des probabilités a l'astronomie* (part of *Traité du calcul des probabilités et de ses applications*) (Paris, 1931)
Christianson, Gale E., *Edwin Hubble: Mariner of the Nebulæ* (New York, 1995)
Clemmensen, C.A., "Jünger, Frederik Gottlieb Edvard", *Dansk biografisk leksikon* 2nd ed. (Copenhagen, 1981)
Clemmensen, C.A., "Christopher Peter Jürgensen", *Dansk biografisk leksikon* 2nd edition, (Copenhagen, 1981)
Corlin, Axel, "The variations with sidereal time in the instensity of highly penetrating cosmic radiation", *Arkiv* vol 21 B no 1(1928)
–"Statistische Behandlung der Höhenstrahlungsmessungen", *Zeitschrift für Physik* vol 50 (1928), 808-848.
–"Measurements of the Cosmic Ultra-Radiation in Northern Sweden", *Lund Observatory Circular* no 6, 1932, 124-132
–*Cosmic Ultra-Radiation in Northern Sweden* (*Annals of the Observatory of Lund*, 4) (Lund, 1934)
–"How larger bodies may be built up out of small particles in interstellar space", *Zeitschrift für Astrophysik*, vol 15 (1938), 239-262
–"On the Building up of Larger Bodies from small Particles in Interstellar Space", *Zeitschrift der Astrophysik* vol 18 (1939), 1-24
"Cosmic", *Encyclopædia Britannica*, 11th ed. (New York, 1910)
Crawford, Elisabeth, *Arrhenius: From Ionic Theory to the Greenhouse Effect* (Canton, MA., 1996)
Crelinstein, Jeffrey, "William Wallace Campbell and the 'Einstein problem': an observational astronomer confronts the theory of relativity", *HSPS* vol 14 (1983), 1-91
Crowe, Michael J., *The Extraterrestrial Life Debate 1750-1900: The Idea of a Plurality of Worlds from Kant to Lowell* (Cambridge, 1986; 1988)

Dahlgren, Thorild, "Sven Dag Wicksell", *Vetenskaps-societetens i Lund årsbok* 1939 (Lund, 1939), 151-161
–"Anteckningar från studentår 1907-1918", in *Under Lundagårds kronor: tredje samlingen* (Lund, 1955), 231-308
Danver, Carl-Gustav, *A Morphological Investigation of Some Near Galaxies with Regard to the Length and the Form of their Arms, their Inclinations and their Symmetry Properties* (*Annals of the Observatory of Lund*, 10) (Lund, 1942)
Daston, Lorraine & Galison, Peter, "The Image of Objectivity", *Representations* vol 40 (1992), 81-128
Davidson, C., "Astrophotographie", in *Handbuch der Wissenschaftlichen und angewandthen Photographie*, vol 6 part 1 (Vienna, 1931), 102-233
DeVorkin, David H., "A Sense of Community in Astrophysics: Adopting a System of Spectral Classification", *Isis* vol 71 (1981)
–"Stellar Evolution and the Origin of the Hertzsprung-Russell Diagram", in Owen Gingerich ed., *The General History of Astronomy* vol 4A, 90-118
Dewhirst, David & Hoskin, Michael, "The Message of Starlight: The Rise of Astrophysics", in Michael Hoskin ed., *The Cambridge Illustrated History of Astronomy* (Cambridge, 1997), 256-343
Dick, Steven J., *The Biological Universe: The Twentieth-Century Extraterrestrial Life Debate and the Limits of Science* (Cambridge, 1996)
Dieke, Sally H., "Heinrich Louis d'Arrest", *DSB*
Doel, Ronald, *Solar System Astronomy in America: Communities, Patronage, and Interdisciplinary Science, 1920-1960* (Cambridge, 1996)
Dunér, Nils & Nordenskiöld, A.E., "Anteckningar till Spetsbergens geografi", *KVAH* vol. 6 no 5 (1865)
–"Förberedande undersökningar rörande utförbarheten af en gradmätning på Spetsbergen", *KVAH* vol. 6 no 8 (1866)
Nils Dunér, *Mesures micrométriques d'étoiles doubles, faites a l'observatoire de Lund, suivies de motes sur leurs mouvements relatifs* (Lund, 1876)
–"Sur les étoiles a spectres de la troisième classe", *KVAH* vol 21 (1884)
–*Några ord angående den föreslagna nya refraktorn å Upsala observatorium* (Uppsala, 1890)
–*Recherches sur la rotation du Soleil* (Uppsala, 1891)
–"Om astronomisk fotografi", *Fotografisk tidskrifts årsbok* 1892, (Stockholm, 1892), 10-16
–*Handbok i allmän astronomi* (Stockholm, 1899)

-"Om den på fotografisk väg framställda stjernkatalogen", *ÖKVAF* 1900, 399-407
-*Über die Rotation der Sonne. Zweite Abhandlung* (*Nova acta regiæ Societatis scientiarum Upsaliensis* ser 4 vol 1 no 6 (Uppsala, 1907)
Eddy, John A., "The Schaeberle 40-ft eclipse camera of Lick observatory", *JHA* vol 2 (1971), 1-22
Ekholm, Nils, "Meteorologiska centralanstalten", *Nordisk familjebok* (Stockholm, 1913)
Ekholm, Nils, & Arrhenius, Svante, "Über den Einfluss des Mondes auf den elektrischen Zustand der Erde. 1-2", *Bihang till k. svenska vetenskapsakademiens handlingar*, vol 19, afd. I no 8 (1894), vol 20 afd. I no 6 (1895)
-"Über den Einfluss des Mondes auf die Polarlichter und Gewitter", *KVAH*, vol 31 no 2 (1898)
-"Über die nahezu 26-tägige Periode der Polarlichter und Gewitter", *KVAH* vol 31 no 3 (1898)
Ekström, Anders, *Den utställda världen: Stockholmsutställningen 1897 och 1800-talets världsutställningar* (*Nordiska muséets handlingar* 119) (Stockholm, 1994)
Eriksson, Gunnar *Kartläggarna: Naturvetenskapens tillväxt och tillämpningar i det industriella genombrottets Sverige 1870-1914* (Umeå, 1978)
Fleck, Ludwik, *Uppkomsten och utvecklingen av ett vetenskapligt faktum: Inledning till läran om tankestil och tankekollektiv* (Stockholm & Stehag, 1997)
Florin, Christina, "Statsbyråkratins erogena zoner: Staten som mötesplats för kvinnligt och manligt", in Gunnar Broberg, Ulla Wikander, Klas Åmark eds., *Svensk historia underifrån: Bryta, bygga, bo* (Stockholm, 1994), 174-198
Fotografisk tidskrifts årsbok 1894-1895 (Stockholm, 1895)
Franzén, Olle, "Lindhagen, Daniel Georg", *SBL*
Friedman, Robert Marc, "Nobelprisen och de vetenskapliga disciplinernas historia: Preliminära tankar och principer", in Gunnar Broberg, Gunnar Eriksson, Karin Johannisson, eds., *Kunskapens trädgårdar: Om institutioner och institutionaliseringar i vetenskapen och livet* (Stockholm, 1988), 136-152
Fries, Thoralf & von Döbeln, Ernst, *Uppsala universitets matrikel höstterminen 1936* (Uppsala, 1937)
Frängsmyr, Tore ed., *Solomon's House Revisited: The Organization and Institutionalization of Science* (Canton, MA, 1990)

Föreläsningar och öfningar vid Stockholms högskola höstterminen 1899 (Stockholm, 1899)
Förteckning över statliga utredningar 1904-1945 utgiven av riksdagsbiblioteket (Norrköping, 1953)
"G. Schiaparelli über die Marstheorie von Svante Arrhenius", *Kosmos* vol 7 (1910), 303
Galison, Peter, *How Experiments End* (Chicago, 1987)
–*Image and Logic: A Material Culture of Microphysics* (Chicago, 1997)
–"Judgment against Objectivity", in Caroline A. Jones & Peter Galison, *Picturing Science, Producing Art* (New York & London, 1998), 327-359
"Gammal klocka sönderslagen", *Arbetet* 29 January 1974
B.P. Gerasimovic, "On Mira Variables and Penetrating Radiation", *Harvard College Observatory Bulletin* no 847 (1927), 1-5
Gingerich, Owen, ed., *Astrophysics and Twentieth-Century Astronomy to 1950* (*The General History of Astronomy* vol 4A) (Cambridge, 1984)
Golinski, Jan, *Making Natural Knowledge: Constructivism and the History of Science* (Cambridge, 1998)
Gooding, David, Pinch, Trevor & Schaffer, Simon, *The Uses of Experiment: Studies in the Natural Sciences* (Cambridge, 1989; 1993)
Gustafsson, Torbjörn, *Själens biologi: Medicinen, kulturen och naturens ordning 1850-1920* (Stockholm & Stehag, 1996)
Gyldén, Hugo, *Framställning af astronomin i dess historiska utveckling och på dess nuvarande ståndpunkt* (Stockholm, 1874)
Gyllenberg, Walter, "Sven Dag Wicksell", *Kungl. Fysiografiska sällskapets i Lund förhandlingar* vol 9 (1939), 40-45
Günter, [review of Arrhenius' Lehrbuch], *Dr. A. Petermanns Mittheilungen aus Justus Perthes' geographischer Anstalt* vol 49 (1903), *Geographischer Literatur-Bericht für 1903*, 152f
Hacking, Ian, *Representing and Intervening: Introductory Topics in the Philosophy of Natural Science* (Cambridge, 1983; 1987)
Handlingar rörande tillsättandet av lediga professorsämbetet i astronomi vid universitetet i Lund 1927-1928 (Lund, 1928)
Handlingar rörande befordringsmål vid Uppsala universitet. Professuren i Uppsala 1938 (Uppsala, 1938)
Hasselberg, Bernhard, "Daniel Georg Lindhagen", *Levnadsteckningar* vol 4 (1899-1912)
–"Om totala solförmörkelser och deras betydelse för studiet af solens fysik, särskildt med hänsyn till den förestående solförmörkelsen den 21 augusti 1914", *Nordisk tidskrift* 1913, 21-40

–[account of the solar eclipse expedition of the Royal Academy of Sciences], *KVAÅ* 1915, 99-125

Hearnshaw, John B., *The Analysis of Starlight: One Hundred and Fifty Years of Astronomical Spectroscopy* (Cambridge, 1986)

–*The Measurement of Starlight: Two Centuries of Astronomical Photometry* (Cambridge, 1996)

Hedin, Marika & Larsson, Ulf, eds., *Teknikens landskap: En teknikhistorisk antologi tillägnad Svante Lindqvist* (Stockholm, 1998)

Hellman, G., [review of Arrhenius' *Lehrbuch*], *Meteorologische Zeitschrift* vol 200 (1903), 527f

Hentschel, Klaus, "Photographic Mapping of the Solar Spectrum 1864-1900, Part I", *JHA* vol 30 (1999), 93-119

Hermodsson, Elisabeth & Trojer, Lena eds., *Ikaros med två bröst: kvinnan och naturvetenskapen* (Stockholm, 1995)

Herrmann, Dieter B., *The History of Astronomy from Herschel to Hertzsprung* (Cambridge, 1984)

–"Hermann Carl Vogel", *DSB*

–"Potsdam Astrophysical Observatory", in Owen Gingerich ed., *Astrophysics and Twentieth-Cenutry Astronomy to 1950* (*The General History of Astronomy* vol 4A) (Cambridge, 1984)

Hetherington, N.S., ed., *Encyclopedia of Cosmology*,

–"Adriaan van Maanen on the Significance of Internal Motions in Spiral Nebulæ", *JHA* vol 5 (1974), 52-53

–"The Simultaneous 'Discovery' of Internal Motions in Spiral Nebulæ", *JHA* vol 6 (1975), 115-145;

–*Science and Objectivity: Episodes in the History of Astronomy* (Iowa City, 1988);

Hetherington, N.S. & Brashear, R.S., "The Hubble-van Maanen Conflict Over Internal Motions in Spiralæ; yet More Information on an Already Old Topic", *Vistas in Astronomy* vol 34 (1991), 415-423

Holmberg, Erik, *A Study of Double and Multiple Galaxies: Together with Inquiries into some General Metagalactic Problems* (Annals of the Observatory of Lund, no 6) (Lund, 1937)

–*Invisible Companions of Parallax Stars Revealed by Means of Modern Trigonometric Parallax Observations* (Lund Medd. Ser II no 92) (Lund, 1938)

–*On the Existence of a Systematic Error in the Estimated Magnitudes of Galaxies* (Lund Medd. ser II no 100) (Lund, 1939)

–*A Photographic Photometry of the Spiral Nebulæ NGC 891 and NGC 3623* (*Lund Medd.* ser II no 114) (Lund, 1945)

–*On the Apparent Diameters and the Orientation in Space of Extragalactic Nebulæ* (Lund Medd. ser II no 117) (Lund, 1946)

–*On the Absorption in the Spiral Nebulæ* (*Lund Medd.* ser II no 120) (Lund, 1947)

–"Lundensisk astronomi under ett sekel", *Cassiopeia: tycho Brahe sällskapets årsbok* 1949, 15-54.

Holmberg, Gustav, "Astronomiskt porträttgalleri", in Gunnar Broberg & Jakob Christensson eds., *Den idéhistoriska bilden* (*Ugglan: Lund Studies in the History of Science and Ideas*, 5) (Lund, 1995)

–"Mechanizing the Astronomer's Vision: On the Role of Photography in Swedish Astronomy, *c.* 1880-1914", *Annals of Science* vol. 53 (1996), 609-616

Hooper, Wynnard, "Statistics", *Encyclopædia Britannica* 11th edition (New York, 1911) vol 25, 806-811

Hoppe, Gunnar, *Till landets gagn: Knut och Alice Wallenbergs stiftelse 1917-1992* (Stockholm, 1993)

Hoskin, Michael A., "Ritchey, Curtis and the Discovery of Novæ in Spiral Nebulæ", *JHA* vol 7 (1976), 47-53

–*The Cambridge Illustrated History of Astronomy* (Cambridge, 1997)

Hubble, Edwin, "Extragalactic Nebulæ", *ApJ* vol 64 (1926), 321-369

Hufbauer, Karl, *Exploring the Sun: Solar Science since Galileo* (Baltimore, 1991)

–"Breakthrough on the Periphery: Bengt Edlén and the Identification of the Coronal Lines, 1939-1945", in Svante Lindqvist ed., *Center on the Periphery: Historical Aspects of 20th-Century Swedish Physics* (Canton, MA, 1993)

–"Artificial Eclipses: Bernard Lyot and the Coronagraph", *HSPS* vol 24 part 2 (1994), 337-394

Internationella fotografiutställningen januari 1934. Vetenskapliga avdelningen Liljevalchs konsthall, catalogue no 102, 45-53

Isaksson, Eva, "Inte en himmelsk syn: kvinnor i Helsingfors Carte du Ciel-projekt, 1893-1930", in Hermodsson, Elisabeth & Trojer, Lena, eds., *Ikaros med två bröst: kvinnan och naturvetenskapen* (Stockholm, 1995)

Jahresbericht am 11. Mai 1877 dem Comité der Nicolai-Hauptsternwarte abgestattet vom Director der Sternwarte (St Petersburg, 1877)

Jahresbericht am 20. Mai 1878 dem Comité der Nicolai-Hauptsternwarte abgestattet vom Director der Sternwarte (St Petersburg, 1878)
Jahresbericht für 1878-79 und 1879-80 am 24. Mai 1880 dem Comité der Nicolai-Hauptsternwarte abgestattet vom Director der Sternwarte (St Petersburg, 1880)
Jahresbericht am 20. Mai 1881 dem Comité der Nicolai-Hauptsternwarte abgestattet vom Director der Sternwarte (St Petersburg, 1881)
Jahresbericht am 19.Mai 1882 dem Comité der Nicolai-Hauptsternwarte abgestattet vom Director der Sternwarte (St Petersburg, 1882)
Jangfeldt, Bengt, *Svenska vägar till S:t Petersburg: kapitel ur historien om svenskarna vid Nevans stränder* (Stockholm, 1998)
Jansson, Essie, *Till minne av Siri* (Stockholm, 1988)
–*Lundaminnen av studentska 1917* (Stockholm, 1989)
Johannisson, Karin, *Det mätbara samhället: Statistik och samhällsdröm i 1700-talets Europa* (Stockholm, 1988)
Johnson, Martin ed., *Knut Lundmark och världsrymdens erövring: en minnesskrift* (Göteborg, 1961)
Jones, Caroline A. & Galison, Peter, *Picturing Science, Producing Art* (New York & London, 1998)
Jonsson, Kjell, *Vid vetandets gräns: Om skiljelinjen mellan naturvetenskap och metafysik i svensk kulturdebatt 1870-1920* (Lund, 1987)
–"Naturvetenskap, världsåskådning och metafysiskt patos i mellankrigstidens Sverige", *Lychnos* 1992, 103-146
–"Physics as Culture: Science and *Weltanschauung* in Inter-War Sweden", in Svante Lindqvist, ed., *Center on the Periphery: Historical Aspects of 20th-Century Swedish Physics* (Canton, MA, 1993)
Jülich, Solveig, "Medicinen och fotografiets mekaniska objektivitet: Carl Curman och tillkomsten av Karolinska institutets fotografiska ateljé 1861", *Lychnos* 1998, 75-120
Kaiserfeld, Thomas, *Vetenskap och karriär: Svenska fysiker som lektorer, akademiker och industriforskare under 1900-talets första hälft* (Lund, 1997)
Kapteyn, J.C., "Statistical Methods in Stellar Astronomy", in *Howard J. Rogers ed., Congress of Arts and Sciences, Universal Exposition, St Louis, Volume IV: Physics, Chemistry, Astronomy, Sciences of the Earth* (Boston and New York 1906), 396-425
Katalog över diapositiv förfärdigade vid Lunds observatorium (Lund, 1929)

Kienle, Hans, "Historical Development of our Ideas Concerning the Structure of the Galaxy", in L.N. Mavridis ed., *Structure and Evolution of the Galaxy* (Dordrecht, 1971), 1-16

Kohler, Robert E., *Lords of the Fly: Drosophila Genetics and the Experimental Life* (Chicago, 1994)

Kristenson, Hjördis, *Vetenskapens byggnader under 1800-talet: Lund och Europa* (Lund, 1990)

Lankford, John, "The Impact of Photography on Astronomy", in Owen Gingerich ed., *Astrophysics and Twentieth-Century Astronomy to 1950* (*The General History of Astronomy* vol 4A) (Cambridge, 1984)

–*American Astronomy: Community, Careers, and Power, 1859-1940* (Chicago, 1997)

Lankford, John & Slavings, Ricky L., "The Industrialization of American Astronomy, 1880-1940", *Physics Today*, January 1996, 34-40

Lightman, Bernard, ed., *Victorian Science in Context* (Chicago, 1997)

Liljequist, Gösta, *High Latitudes: A History of Swedish Polar Travels and Research* (Stockholm, 1993)

Lindblad, Bertil, "On the Use of Grating Spectra for Determining Spectral Type and Absolute Magnitude of the Stars" *ApJ* vol. 49 (1919), 289-302

–"Note to Dr. Knut Lundmark and Dr. W.J. Luyten's Paper on the Determination of the Colour-Equivalent of a Star", *MNRAS* vol 83 (1922-1923), 97-98

–"Spectrophotometric Methods for Determining Stellar Luminosity", *ApJ* vol 55 (1922), 85-118

–"De spektroskopiska metoderna för bestämning av stjärnornas avstånd", *PAT* vol 3 (1922), 33-42

–"On the Intensity-Distribution in Short Grating Spectra and Objective-Prism Spectra s a Function of Spectral Type and Absolute Magnitude", *MNRAS* vol 83 (1923), 503-510

–"On the Cause of Star-Streaming", *ApJ* vol 62 (1925), 191-197

–"Spektralfotometriska undersökningar vid Upsala observatorium", *NAT* ns vol 7 (1926), 41-50

–"Kosmogoniska problem i samband med nyare föreställningar om stjärnsystemets natur", *PAT* vol 7 (1926), 125-134

–"On the Dynamics of the Stellar System", *Arkiv* vol 21A no 3 (1928)

–"On the Development of Spiral Structure in Anagalactic Nebulæ", *Arkiv* vol 22A no 11 (1930)

–"Våra dagars astronomiska huvudproblem. Intryck från den Internationella Astronomiska Unionens kongress i Cambridge, U.S.A. i september 1932", *PAT* vol 13 (1932), 163-178
–"On the Evolution of a Rotating System of Material Particles, with Applications to Saturn's Rings, the Planetary System and the Galaxy", *MNRAS* vol 94 (1934), 231-240
–"Varifrån och varthän går utvecklingen?", in *Olösta gåtor och aktuella problem: sjuttiosex uttalanden i Svenska Dagbladets enquête om naturforskningens nuvarande uppgifter och mål* (Stockholm, 1937), 157-161
–"Spectral Types and Distribution of Special object", *Transactions of the International Astronomical Union* vol 6 (Cambridge, 1939), 454-457
–"Adress", *Transactions of the International Astronomical Union* vol 6 (Cambridge, 1939), 313f
– *On the Distribution of Light-Intensity and Colour in the Spiral Nebula NGC 7331* (Stock. Ann) vol 13 no 8 (1941)
–"Gyldén, Johan August Hugo", *SBL*
–"Östen Bergstrand", *Levnadsteckningar* vol. 9, 489-499
Lindblad, Bertil & Stenquist, Erik, *On the Spectrophotometric Criteria of Stellar Luminosity* (*Astronomiska iakttagelser och undersökningar å Stockholms observatorium* vol. 11 no 12) (Stockholm & Uppsala, 1934)
Linders, Frans Josua, *Zur Kenntnis der Kopfmasse in Schweden* (*Lund Medd.* Ser II no 50A) (Lund, 1927) (also in *Festskrift tillägnad C.V.L. Charlier på hans sextiofemårsdag den 1 april 1927* (Lund, 1927))
Lindqvist, Svante, "Ett experiment år 1744 rörande norrskenets natur", in Gunnar Broberg, Gunnar Eriksson, Karin Johannisson eds., *Kunskapens trädgårdar: Om institutioner och institutionaliseringar i vetenskapen och livet* (Stockholm, 1988)
–"A Wagnerian Theme in the History of Science: Scientific Glassblowing and the Role of Instrumentation", in Tore Frängsmyr ed., *Solomon's House Revisited: The Organization and Institutionalization of Science* (Canton, MA, 1990), 160-183
– ed., *Center on the Periphery: Historical Aspects of 20th-Century Swedish Physics* (Canton, MA, 1993)
Lund observatory yearly reports, *LUÅ*
Lundahorizonten 1938, 1938
Lunds universitets matrikel läsåret 1924-1925 (Lund, 1925)
Lundell, J.A., "De akademiska sommarkurserna i Uppsala", *Verdandi* 1894

Lundborg, Hermann & Linders, Frans Josua, *The Racial Characters of the Swedish Nation* (Uppsala, 1926)
Lundgren, Anders, "Discipliner och institutionalisering inom 1900-talets svenska vetenskaper", in Bo Sundin ed., *Från hermetism till rationell distribution: Föredrag vid svensk idéhistorisk konferens 1992* (Umeå, 1993), 99-110
–*Vetenskap till vardags: En historia kring Xylocain* (Uppsala, 1995)
Lundmark, Knut, "Bestimmung der Bahn des Kometen 1802", *Arkiv för matematik, fysik och astronomy* vol 12 (1917), no 17
–"Definitive Bestimmung der Bahn des Kometen 1802", *AN* vol 202 (1917)
–"Mitteilung über die Nova Aquilæ 3", *AN* no 4950 vol 207 (1918)
–"Die Stellung der kugelförmigen Sternhaufen und Spiralnebel zu unseren Sternsystem: Ein Versuch, ihre Parallaxen zu schätzen", *AN* vol 209 (1919), 369-380
–"The Relations of the Globular Clusters and Spiral Nebulæ to the Stellar System: An Attempt to Estimate their Parallaxes", *KVAH* vol 60 no 8 (1920)
–"The Spiral Nebula Messier 33", *PASP* vol 33 (1921), 324-327
–*Spiralnebulosorna: Ett nutida astronomiskt forskningsproblem* (Studentföreningen Verdandis småskrifter no 247) (Stockholm, 1922)
–"On the Motions of Spirals", *PASP* vol 34 (1922), 108-115
–"The Distance of the Large Magellanic Cloud", *Observatory* vol 47 (1924), 276-279
–"The Motions and the Distances of Spiral Nebulæ", *MNRAS* vol 85 (1925), 865-894
–"Den internationella astronomiska unionen", *PAT* vol 6 (1925), 169-173, 171
–"Om spiralnebulosornas indelning, dimensioner och utveckling", *PAT* vol 6 (1925), 104-114.
–"A Preliminary Classification of Nebulæ", *Arkiv* vol 9B (1926), no 8
–"Internal Motions of Messier 33", *ApJ* vol 63 (1926), 67-71
–*Världsrymdens liv* (Uppsala, 1926)
–"Studies of Anagalactic Nebulæ. First Paper", *Nova Acta RegiæSocietatis Scientiarum Upsaliensis.* Volumen Extra Ordinem Editum 1927 (Uppsala, 1927)
–"Ignorabimus? Till frågan om världens oändlighet", *Religion och kultur* vol 1 (1930), 89-100.
–"A New General Catalogue of Nebulæ", *PASP* vol 42 (1930), 31-33

-"Kurser i vetenskaplig fotografi vid Lunds universitet", *Nordisk tidskrift för fotografi*, 1934 no 6, 77-80
-*Livets välde: Till frågan om världarnas beboelighet. I.* (Stockholm, 1935)
-*Astronomien i Lund 1667-1936: Glimtar och drag ur astronomiska institutionens liv och öden i nuvarande och gången tid* (Lund, 1937)
-"Enhet och mångfald i universum", in *Olösta gåtor och aktuella problem: sjutiosex uttalanden i Svenska Dagbladets enquête om naturforskningens nuvarande uppgifter och mål* (Stockholm, 1937)
-"Backlund, Johan Oskar (Oskar Andreevitj)", *SBL*
-"Carl Östen Emanuel Bergstrand", *SBL*
-"Charlier, Carl Vilhelm Ludvig", *SBL*
-"Wicksell, Sven Dag", *SMoK*
Lundmark, Knut, & Lindblad, Bertil, "Photographic Effective Wavelengths of Nebulæ and Clusters, Second Paper", *ApJ* vol 50 (1919), 376-390
Lundmark, Knut, & Luyten, W.J., "On the Determination of the Colour-Equivalent of a Star, with Special Reference to the Effective Wave-Length and its Relation to Spectral Class: A Review of the Different Methods", *MNRAS* vol 82 (1922), 495-509
-"Note on the Determination of Absolute Magnitude from λ_e and λ_m", *MNRAS* vol 83 (1923), 470-474
Lundmark, Knut & Hubble, Edwin, "Nova Z Centauri", *PASP* vol 34 (1922), 292-293
Lunds universitets matrikel läsåret 1924-25 (Lund, 1925)
Lönnqvist, Conrad, "A Criticism of the Hypothesis of Electrostatic Accumulation of Meteorites", *Zeitschrift für Astrophysik* vol 17 (1939), 344-359
Malmquist, Gunnar, "Några ord om avståndsenheterna inom stellarastronomien", *PAT* vol 4 (1925), 46-48
-"Bestämning av stjärnornas färg i större skala med Seares' metod", *PAT* vol 5 (1924), 19-25
-"Hamburgs observatorium i Bergedorf", *PAT* vol 6 (1925), 35-43
- "Einige Bemerkungen über die räumliche Verteilung der Sterne", *Vierteljahrsschrift der Astronomischen Gesellschaft* vol 61 (1926), 244-250
-"On the Determination of Colour Indices with Reflectors", *Stock. Ann.* vol 13 no 1 (1938)
- [no title], *Transactions of the International Astronomical Union* vol 6 (Cambridge, 1939), 459-461

–"Hur Uppsala fick sitt nuvarande observatorium: ett hundraårsminne", *PAT* vol 35 (1954)
–"Carl Vilhelm Ludvig Charlier", *KVAÅ* vol 58 (1960), 385-405
Malmquist, Sten, "Linders, Frans Josua", *SBL*
Masloff, Inez, "Knut Lundmark i Åbo", in Johnson ed., *Knut Lundmark*
Mavridis, L.N., ed., *Structure and Evolution of the Galaxy* (Dordrecht, 1971), 1-16
Meijer, Hans, *Kommittépolitik och kommittéarbete: Det statliga kommittéväsendets utvecklingslinjer 1905-1954 samt nuvarande funktion och arbetsformer* (Lund, 1956)
Mihalas, Dimitri, *Galactic Astronomy* (San Francisco, 1968)
Mitchell, S.A., "Comet 1903 Borrelly and light-pressure", *ApJ* vol 20 (1904)
Moustgaard, Lisbeth, "Siriometer, siriusvidde og parsec. Enheder for afstande til stjernerne - den historiske udvikling", *AT* vol 24 (1995), 14-19
Müller, Johannes, *Lehrbuch der kosmischen Physik* (Müller & Pouillet, *Lehrbuch der Physik und Meteorologie* volume three) (Braunschweig, 1856)
Nelson, Axel, *Akademiska afhandlingar vid Sveriges universitet och högskolor läsåren 1890/91-1909/10 jämte förteckning öfver svenskars akademiska afhandlingar vid utländska universitet under samma tid* (Uppsala, 1911)
Newcomb, Simon & Engelmann, Rudolf, *Populäre Astronomie* 6th ed (Leipzig, 1921)
Nielsen, Axel, "Contributions to the History of the Hertzsprung-Russell Diagram", *Centaurus* vol 9 (1964), 219-253
Nilson, Peter, "Catalogues of Galaxies. A Short Survey of their History", in Peter Nilson, *Uppsala General Catalogue of Galaxies* (*Nova Acta Regiæ Societatis Scientiarum Upsaliensis* ser. V:A vol 1 = *Uppsala observatoriums annaler* vol 6) (Uppsala, 1973)
Nordenmark, N.V.E., "Stjärneborgsproblemets historia under femtio år", *PAT* vol 32 (1951), 87-93
–"Kring Tycho Brahe-jubileet 1901", *PAT* vol 33 (1952), 15-22
Norlind, Wilhelm & Olsson, Bror, *Lunds universitets matrikel 1939* (Lund, 1940)
"Nova Aquilæ 3", *AN* no 4949 vol 207 (1918)

Nybom, Thorsten, "Samhällsformation och samhällsorganisation i Sverige 1890-1975 – en principskiss", in Thorsten Nybom & Rolf Torstendahl, eds., *Byråkratisering och maktfördelning* (Lund, 1989)

Nyrén, Magnus, "Johan Oskar (Oskar Andreevitsch) Backlund", *Levnadsteckningar* vol 5 (1915-1920)

"Observatorieuret sönderslaget", *Skånska Dagbladet* 29 January 1974

Ohlon, Sven, "Hasselberg, Clas Bernhard", *SBL*

Olösta gåtor och aktuella problem: sjuttiosex uttalanden i Svenska Dagbladets enquête om naturforskningens nuvarande uppgifter och mål (Stockholm, 1937)

Oort, Jan H., "Obituary Notices: Bertil Lindblad", *QJRAS* vol 7 (1966), 329-341

Oredsson, Sverker, *Lunds universitet under andra världskriget: Motsättningar, debatter och hjälpinsatser* (*Lunds universitetshistoriska sällskap årsbok* 1996) (Lund, 1996)

Osterbrock, Donald, *James E. Keeler: Pioneer American Astrophysicist and the Early Development of American Astrophysics* (Cambridge, 1984)

–"Getting the Picture: Wide-Field Astronomical Photography from Barnard to the Achromatic Schmidt, 1888-1992", *JHA* vol 25 (1994), 1-14

Osterbrock, Donald, Gustafson, John R. & Unruh, W.J. Shiloh, *Eye on the Sky: Lick Observatory's first Century* (Berkeley, Los Angeles, & London, 1988)

Paul, Erich Robert, "Kapteyn and Statistical Astronomy", in Hugo van Woerden, Ronald J. Allen & W. Butler Burton eds., *The Milky Way Galaxy* (Dordrecht, 1985), 25-42

–*The Milky Way Galaxy and Statistical Cosmology, 1890-1924* (Cambridge, 1993)

Pang, Alex Soojung-Kim, "The Social Event of the Season: Solar Eclipse Expeditions and Victorian Culture", *Isis* vol 84 (1993), 252-277

–"Victorian Observing Practices, Printing Technology, and Representations of the Solar Corona, (1): The 1860s and 1870s", *JHA* vol 25 (1994), 249-274

–"Victorian Observing Practices, Printing Technology, and Representations of the Solar Corona, (2): The Age of Photomechanical Reproduction", *JHA* vol 26 (1995), 63-75

Persson, Bodil, *När kvinnorna kom in i männens värld: Framväxten av ett kvinnligt yrke. Laboratorieassistenter under perioden 1880-1941* (Stockholm, 1994)

Petrén, Gustaf, "Minnen och interiörer rörande professor Charliers installation i Lund år 1897", *Under Lundagårds kronor: tredje samlingen* (Lund, 1955), 95-109
Pickering, Andrew, *The Mangle of Practice: Time, Agency, and Science* (Chicago, 1995)
Ramberg, Jöran, "Ljusfördelningen i solkoronan: Resultat från de totala solförmörkelserna 1945 och 1914", *PAT* vol 32 (1951), 100-120
–"Astronomiska studier i Sydafrika", *PAT* vol 33 (1952), 122-142
Rogers, Howard J., ed., *Congress of Arts and Sciences, Universal Exposition, St Louis, Volume IV: Physics, Chemistry, Astronomy, Sciences of the Earth* (Boston and New York 1906)
Runeby, Nils, "Americanism, Taylorism and Social Integration", *Scandinavian Journal of History* vol 3 (1978), 21-46
"Rättaste klockan i Lund nu lagad", *Arbetet* 11 January 1975"
Schaffer, Simon, "Where Experiments End: Tabletop Trials in Victorian Astronomy", in Jed Z. Buchwald ed., *Scientific Practice: Theories and Stories of Doing Physics* (Chicago & London, 1995), 257-299
Schalén, Carl, "Spectral Classification of Faint Milky Way Stars in Aquila", *Arkiv* vol 18 no 36 (1925)
–"Spectro-photometric Determinations of Absolute magnitudes of B and A Type Stars", *Arkiv* vol 19A no 33 (1926)
–"The Space Distribution of B and A Type Stars in Bright and Dark Galactic Regions. Researches Based on Determinations of Spectral Type and Luminosity Class for 4300 Stars", *KVAH* ser 3 vol 6 no 6 (1928)
–"Om en allmän absorption av ljuset i världsrymden", *Beretn. om det 18. Skand. Naturforskermöde i Köbenhavn* (Köpenhamn, 1929)
–"Zur Frage einer allgemeinen Absorption des Lichtes im Weltraum", *AN* no 5656 vol 236 (1929), 249-258
–"Om förekomsten av mörk materie i världsrymden", *PAT* vol 11 (1930), 22-44
–"Untersuchungen über Dunkelnebel", *KVAH* ser 3 vol 13 no 2 (1934)
–"Über Probleme der interstellaren Absorption", *Nova acta R. Soc. Scient. Ups.* Ser 4 vol 10 (1936)
"Schalén, Carl Adam Wilhelm", *Lunds universitets matrikel 1967-1968*
Schalén, Carl, Hansson, Nils, Leide, Arvid, *Astronomiska observatoriet vid Lunds universitet* (*Ur Lunds universitets historia*, 4) (Lund, 1968)
Schalén, Carl & Lindblad, Bertil, "The Luminosities, Individual Parallaxes and Motions of B and A Type Stars", *Arkiv* vol 20A no 7 (1927)

Schwarzschild, Karl, "Der Druck des Lichts auf kleine Kugeln und die Arrhenius'sche Theorie der Cometenschweife", *Sitzungsberichte der math.-phys. Classe der k.b. Akademie der Wissenschaften zu München* vol 31 (1901)
Sekido, Yataro & Elliot, Harry, *Early History of Cosmic Ray Studies* (Dordrecht, 1985)
Shapley, Harlow, "On the Classification of Extra-Galactic Nebulæ", *Harvard College Observatory Bulletin* no. 849 (1927)
Shapley, Harlow & Ames, Adelaide, *A Survey of the External Galaxies Brighter than the Thirteenth Magnitude* (*Annals of the Harvard College Observatory* vol 88, no 2) (Cambridge, Mass., 1932)
Silverbark, Thord, *Fysikens filosofi: Diskussioner om Einstein, relativitetsteorin och kvantfysiken i Sverige 1910-1970* (Stockholm & Stehag, 1999)
Skoglund, Crister, *Vita mössor under röda fanor: Vänsterstudenter, kulturradikalism och bildningsideal i Sverige 1880-1940* (Stockholm, 1991)
Skottsberg, C., ed., *Kungl. Svenska Vetenskapsakademien: Personförteckningar 1916-1955* (Stockholm, 1957)
Smith, Robert W., *The Expanding Universe: Astronomy's 'Great Debate' 1900-1931* (Cambridge, 1982)
–"Studies of the Milky Way 1850-1930: Some Highlights", in van Woerden, Hugo, Allen, Ronald J., Burton, W. Butler, eds., *The Milky Way* (IAU Symposium no 106) (Dordrecht, 1985), 43-58
–"Engines of Discovery: Scientific Instruments and the History of Astronomy and Planetary Science in the United States in the Twentieth Century", *JHA* vol 28 (1997), 49-77.
–"Cosmology 1900-1931", Hetherington ed., *Encyclopedia of Cosmology*, 116-126
Sommarkurserna i Uppsala 1895: Förteckning öfver deltagarne (Uppsala, 1895)
Sommarkurserna i Uppsala 1895: Tillägg och rättelser till förteckning öfver deltagarne (Uppsala, 1895)
"Sommarkurserna i Uppsala", *Svensk läraretidning* vol 12 no 35
Stawström, Carl-Olow, "Relative Acceptance: The Introduction and Reception of Einstein's Theories in Sweden, 1905-1965", in Lindqvist, Svante, ed., *Center on the Periphery: Historical Aspects of 20th-Century Swedish Physics* (Canton, MA, 1993)
Steinmaurer, Rudolf, "Erinnerungen an V.F. Hess, den Entdecker der Kosmischen Strahlung, und an die ersten Jahre des Betriebes des Hafe-

lekar-Labors", in Sekido, Yataro & Elliot, Harry, *Early History of Cosmic Ray Studies* (Dordrecht, 1985)
Stenquist, Erik, *A Spectrophotometric Study of The Cambridge Proper Motion Stars* (*Nova acta R. Soc. Scient. Ups.* Ser 4 vol 10) (Uppsala, 1937)
Stockholm observatory yearly reports, *ÖKVAF* and *KVAÅ*
"Stroh, Alfred Henry", *SMoK*
Strömgren, Elis & Strömgren, Bengt, *Lærebog i Astronomi paa Grundlag af H. Geelmuydens lærebog* (Oslo, 1931)
"Studieplan i matematisk statistik vid Lunds universitet 1917-1918", *Svenska aktuarieföreningens tidskrift* vol 4 (1917), 187f
Sundin, Bo, ed., *Från hermetism till rationell distribution: Föredrag vid svensk idéhistorisk konferens 1992* (Umeå, 1993)
Sundman, Anita, *Den befriade himlen: Ett porträtt av Knut Lundmark* (Stockholm, 1988)
–"Nordenmark, Nils Victor Emanuel", *SBL*
Svenonius, Björn, *Contributions to the Photographic Study of Colour, Resolution, and Total Magnitude of Anagalactic Objects with Special Regard to Objects Photographed at Helwan* (*Annals of the Observatory of Lund* no 7) (Lund, 1938)
Swedenborg, Emanuel, *Opera quaedam aut inedita aut obsoleta de rebus naturalibus*, I-III (Stockholm, 1907-1911)
Söderberg, Rolf & Rittsel, Pär, *Den svenska fotografins historia* (Stockholm, 1983)
Teerikorpo, Pekka, "Lundmark's unpublished 1922 nebula classification", *JHA* vol 20 (1989), 165-170
Transactions of the International Astronomical Union vols 2-6
Tuneld, John, *Akademiska avhandlingar vid Sveriges universitet och högskolor läsåren 1910/11-1939/40* (Lund, 1945)
Turner, Frank M., "Practicing Science: An Introduction", in Bernard Lightman ed., *Victorian Science in Context* (Chicago, 1997)
Under Lundagårds kronor: tredje samlingen (Lund, 1955)
Uppsala observatory yearly report, *UUÅ*
Van Helden, Albert & Hankins, Thomas L., *Instruments* (*Osiris* 2nd ser vol 9) (Chicago, 1994)
–"Introduction: Instruments in the History of Science", in idem, *Instruments*, 1-6
Veckojournalen no 10 1935

Vaucouleurs, Gerard de, Hansson, Nils & Lyngå, Gösta, "Dunér's Observations of SN 1885 (S Andromedæ) in M31", *PASP* vol 97 (1985), 30-31
Wallén, Axel, "Nils Gustaf Ekholm", *KVAÅ* 1925, 273-285
Wallenquist, Åke, "A Photometric Research on the Galactic Cluster Messier 11 (N.G.C. 6705)", *Nova Acta Regiæ Societatis Scientiarum Upsaliensis* ser 4 vol 10 no 4 (1936)
–"Some Preliminary Results on the Space Distribution of White Stars in the Direction of the Galactic Centre", *Arkiv* vol 25A no 27 (1937)
–"On the Dark Nebula at theta Ophiuchi", *Arkiv* vol 25B no 28 (1937)
–"Nyrén, Magnus", *SMoK*
–"Schalén, Carl Adam Wilhelm", *SMoK*
Warwick, Andrew, "The Laboratory of Theory or What's Exact about the Exact Sciences?", in M.Norton Wise ed., *The Values of Precision* (Princeton, 1995)
Wicksell, Sven, "Individens fruktsamhet och släktets förökning", *Ekonomisk tidskrift* 1930
"Wicksell, Sven Dag", *Lunds universitets matrikel läsåret 1924-1925* (Lund, 1925)
–*Mellan kartan och verkligheten: Geodesi och kartläggning, 1695-1860* (Uppsala, 1990)
–"Porträtt av vetenskapsmän", Gunnar Broberg ed., *Gyllene äpplen: Svensk idéhistorisk läsebok* (Stockholm, 1991), 840-842
–"Science and Neutrality: The Nobel Prizes of 1919 and Scientific Internationalism in Sweden", *Minerva* vol. 33 (1995), 339-360
Wise, M. Norton ed., *The Values of Precision* (Princeton, 1995)
Woerden, Hugo van, Allen, Ronald J., & Burton, W. Butler, eds., *The Milky Way Galaxy* (Dordrecht, 1985)
Wolf, Rudolf, *Handbuch der Astronomie ihrer Geschichte und Litteratur* (Zürich 1890-1893)
Wolfschmidt, Gudrun, *Milchstrasse, Nebel, Galaxien: Strukturen in Kosmos von Herschel bis Hubble* (München, 1995)
Wråkberg, Urban, *Vetenskapens vikingatåg: perspektiv på svensk polarforskning 1860-1930* (Stockholm, 1995)
von Zeipel, Hugo, "Nils Christofer Dunér", *KVAÅ* 1916, 290-312
–"Några önskemål för svensk astronomisk forskning", *PAT* 1922, 62-64
Ångström, Anders, "Ekholm, Nils Gustaf", *SBL*
"O.A. Åkesson" in *Svensk uppslagsbok* second edition (1947-1955)

Öhman, Yngve, "Photographic Density-Diagrams of Short Stellar Spectra", *Arkiv* vol 19B no 6 (1926)
–"Photometric Studies of Effects of Luminosity and Colour in Short Stellar Spectra", *Arkiv* vol 20A no 23 (1927).
– *Spectrophotometric Studies of B, A and F Type Stars* (*Nova Acta Regiæ Societatis Scientiarum Upsaliensis* ser. 4 vol 7 no 3) (Uppsala, 1930)
–"Note on the Spectrum of a Probable White Dwarf in h Persei", *MNRAS* vol 92 (1931), 71-72
–"En kombination av reflektor och refraktor", *PAT* vol 12 (1931) 161f
–"Spectrographic Studies in the Red", *ApJ* vol 80 (1934), 171-180
–"Effects of Polarisation in the Spectrum of ß Lyræ", *Nature* vol 134 (1934) 534-536
–*Some Preliminary Results from a Study of Hydogen Absorption for Stars in h and χ Persei*, *Stock. Ann.* vol 12 no 1 (1935)
–"The Red Spectral Region of Dwarf Stars of Class M", *Stock. Ann.* vol 12 no 3 (1936)
–"On the Bands of Magnesium Hydride in Stellar Spectra", *Stock. Ann.* vol 12 no 8 (1936)
–"A New Monochromator", *Nature* vol 141 (1938), 157-160
–"A New Monochromator", *Nature* vol 141 (1938), 291
–"Lindblad, Bertil", *SBL*

INDEX